NASA AND THE LONG CIVIL RIGHTS MOVEMENT

UNIVERSITY PRESS OF FLORIDA

Florida A&M University, Tallahassee
Florida Atlantic University, Boca Raton
Florida Gulf Coast University, Ft. Myers
Florida International University, Miami
Florida State University, Tallahassee
New College of Florida, Sarasota
University of Central Florida, Orlando
University of Florida, Gainesville
University of North Florida, Jacksonville
University of South Florida, Tampa
University of West Florida, Pensacola

NASA

AND THE LONG
CIVIL RIGHTS MOVEMENT

EDITED BY

Brian C. Odom and Stephen P. Waring

University Press of Florida

Gainesville · Tallahassee · Tampa · Boca Raton

Pensacola · Orlando · Miami · Jacksonville · Ft. Myers · Sarasota

University Press of Florida, 2019
Published in the United States of America.

24   23   22   21   20   19   6   5   4   3   2   1

Library of Congress Cataloging-in-Publication Data
Names: Odom, Brian C., editor. | Waring, Stephen P., editor.
Title: NASA and the long civil rights movement / edited by Brian C. Odom and
Stephen P. Waring.
Description: Gainesville : University Press of Florida, 2019. | Includes
bibliographical references and index. |
Identifiers: LCCN 2019013057 (print) | LCCN 2019017023 (ebook) | ISBN
9780813057323 (ePDF) | ISBN 9780813066202 (cloth : alk. paper)
Subjects: LCSH: United States. National Aeronautics and Space
Administration—Appropriations and expenditures. | Civil rights
movements—United States. | African American astronauts. | Black lives
matter movement. | Poverty—United States.
Classification: LCC TL521.312 (ebook) | LCC TL521.312 .N363 2019 (print) |
DDC 629.40973/09046—dc23
LC record available at https://lccn.loc.gov/2019013057

The University Press of Florida is the scholarly publishing agency for the State University System
of Florida, comprising Florida A&M University, Florida Atlantic University, Florida Gulf Coast
University, Florida International University, Florida State University, New College of Florida,
University of Central Florida, University of Florida, University of North Florida, University of
South Florida, and University of West Florida.

University Press of Florida
2046 NE Waldo Road
Suite 2100
Gainesville, FL 32609
http://upress.ufl.edu

CONTENTS

FIGURES

FOREWORD

How We Tell about the Civil Rights Movement and Why It Matters

JACQUELYN DOWD HALL

Few public figures today would dream of denying that "the civil rights movement was a good thing."[1] But plenty would, implicitly or explicitly, posit a sharp break between the "good" civil rights movement, which overcame real evils in the distant past, and current campaigns such as Black Lives Matter or the Moral Monday Movement in North Carolina. That invidious distinction between then and now rests on a sanitized "Montgomery-to-Memphis" narrative about a short, triumphant southern civil rights movement led by a few individuals who relied on restrained, dignified strategies and pursued limited goals. In this version of the past, the movement started with the Montgomery bus boycott of 1955, came to a climax with the passage of the Civil Rights and Voting Rights Acts, and ended with Martin Luther King Jr.'s assassination in 1968. Its primary aim was to overturn the South's explicitly race-based segregation and disfranchisement laws, and it stood in sharp contrast to both Black Power and northern struggles.

In a presidential address I delivered to the Organization of American Historians in 2004 and an article titled "The Long Civil Rights Movement and the Political Uses of the Past" I published the next year, I tried to embed the heroic struggles of the 1950s and 1960s in the context of what I had come to understand as a longer, broader, deeper narrative.[2] I pointed to the modern civil rights era's beginnings in the late 1930s and early 1940s. This was the period in which African American activism, which took place primarily on the local level, became national in scope.

The collapse of the South's sharecropping system, black migration to the cities, the emergence of a strong federal government, critical legal victories on the part of the National Association for the Advancement of Colored People (NAACP), the example of anticolonial uprisings, the opportunities offered to black workers by an insurgent labor movement, the efforts of blacks and whites on the left edge of the New Deal coalition to build an egalitarian welfare state and put racial equality on the national political agenda—all these enabled local civil rights struggles to unfold in a vastly transformed environment.

Picking up on the work of Robert Korstad, Nelson Lichtenstein, and others, I suggested that the Cold War and McCarthyism rolled back the unions and stigmatized the Left, helping to route the civil rights movement into new channels and erasing the black activists who pursued what Korstad called "civil rights unionism" from popular memory.[3] I characterized the upheavals of the 1950s, 1960s, and 1970s as a "movement of movements" rather than as series of mutually exclusive battles.[4] I tried to direct attention to the struggles for access to schools and jobs that continued and in some cases intensified after the fall of de jure segregation. Finally, drawing on a burgeoning new historiography on conservatism, I underscored the point that the long civil rights movement flowed in parallel with a long conservative movement that metamorphosed over time as it resisted and, at critical moments, thwarted civil rights gains.

By considering the progressive and conservative movements of the mid-twentieth century in tandem, I came to see that the story of a short, successful civil rights movement has persisted in part because it served the goals of those who, like the architects of the recent wave of voter suppression laws, want to wrap themselves in the mantle of the movement even as they undo many of its gains. Conservatives in the 1930s and 1940s opposed the expansion of the New Deal and fought tooth-and-nail against advocates of unionization and black civil rights. In the 1970s, however, they reinvented themselves as the guardians of color-blind laws, which they took to be the civil rights movement's singular goal. Since that goal has been attained, they suggest, individual African Americans bear the onus of their own failure or success. If stark group inequalities persist, whether in wealth or political power or treatment by the criminal justice system, then blacks' attitudes, behavior, and family structures are to blame. Therefore, the thinking goes, we do not need group remedies such as public employees' unions to protect workers in the sectors where minorities and women have made the most progress, or color- and class-conscious policies aimed at creating integrated schools, or affirmative efforts to open educational and employment opportunities to traditionally excluded groups, or the oversight mandated by the Voting Rights Act.

When I began writing about these matters fifteen years ago, I had the exciting sense of having my finger on the pulse of a vital scholarly trend.[5] Since then so many scholars in different fields and working on different periods and topics have augmented the "long civil rights movement" theme that an adequate accounting would require its own historiographical essay.[6] *NASA and the Long Civil Rights Movement* offers yet another intriguing perspective on what has rightly been called America's "original sin" and on efforts to redeem our country from its foundations in slavery, segregation, and racial violence. Margot Lee Shetterly's *Hidden Figures*, along with the movie based on the book, brought welcome attention to the black women mathematicians at NASA who, despite their treatment as second-class citizens, made heroic contributions to the Space Race that, in tandem with the civil rights and women's movements, so powerfully shaped mid-twentieth-century American life.[7] In Brian C. Odom and Stephen P. Waring's edited volume, we see how, during this drive for aeronautical advancement, African Americans tried to turn the power of the federal government to their advantage, how white southern segregationists simultaneously argued for states' rights and took advantage of federal largesse and how the space program served as an agent of modernization and racial moderation, while at the same time displacing and excluding African Americans and promoting new, ostensibly gender- and color-blind disparities.

I have learned a tremendous amount from all of this new work. Among other things, I have had to think harder about the continuities and discontinuities between the New Deal generation and my own, which came to political consciousness in the 1960s. I originally stressed the degree to which McCarthyism silenced the black and white activists of the 1930s and 1940s and thus narrowed the range of acceptable political discourse on which the civil rights movement could draw. My current work on that earlier generation, along with new research by others, has reinforced my conviction that it is vitally important to remember the damage that anti-Communist hysteria did and that similar forms of hysteria can still do.[8] Yet I am also persuaded by recent scholarship that I downplayed the links between the New Deal era and the later period and so underestimated the continuity of, to take two examples, black left feminism and job-related struggles.[9] And I take the point of those who advocate a longer, broader, deeper narrative but ground that narrative in studies of local organizing traditions in the rural South.[10]

A small group of scholars remains invested in drawing sharp distinctions between time periods, regions, strategies, and ideologies, especially between the social democratic activism of the 1930s and 1940s and that of the Montgomery-to-Memphis movement, between nonviolence and Black Power, and between the South and the North.[11] I certainly understand these concerns: it is always important to account for difference and discontinuity, even as the passage of time and the accumulation of research allow us to take an ever longer and broader view. I am even more sympathetic to the perspective articulated by Kevin Boyle, who wants to push beyond both the traditional and long civil rights narratives to one centered on "a long, slow slog of struggle against deeply entrenched social, economic, and political structures that, while diminished, have yet to give way."[12] Still, I would stand by the value of the periodization I proposed, with the proviso that I see all periodization as a framework we impose in order to make sense of the messy reality of the past. And I am less interested in historians' quibbles about how to do that slicing and dicing than in what Jeanne Theoharis terms "the uses and misuses" of the past.[13]

A recent encounter with a dedicated public-school teacher reminded me, once again, of the urgency of our choices about which stories to tell and of how complicated popular memory can be. In response to a talk I gave at a seminar sponsored by the Southern Oral History Program at the University of North Carolina, Hugh Davis, a librarian and English teacher in Winston-Salem, North Carolina, reflected on "the effect of bottling up the civil rights movement" so that it "covers just a few years, with a few key figures, and . . . culminated in the passage of the Voting Rights Act." His students, virtually all of whom are black, "watch the news every day to see another story of an African American killed in the streets or subject to police brutality." Given that reality, they (unlike the white color-blind conservatives I wrote about) "see the movement as a failure." To Hugh Davis, viewing "the civil rights movement in a larger context . . . produces hope that we might be in dark times, but that does not mean it will always stay that way." By reframing the movement in that way, "I can discuss with my students the fact that times may be tough, but there are peaks and valleys of human behavior. The tough times they see are just part of a valley. That mountaintop (to co-opt Dr. King's metaphor) is still within sight."[14]

Countless efforts to reach that mountaintop are under way as we speak. Not least among them are books like this one: self-critical and aimed at strengthening key governmental institutions by confronting their past failures and their efforts to live up to their most lofty missions.

Notes

1. J. Todd Moye, *Let the People Decide: Black Freedom and White Resistance Movements in Sunflower County, Mississippi, 1945–1986* (Chapel Hill: University of North Carolina Press, 2004), 214.

2. Jacquelyn Dowd Hall, "The Long Civil Rights Movement and the Political Uses of the Past," *Journal of American History* 91, no. 4 (2005): 1233–1263.

3. Robert Korstad and Nelson Lichtenstein, "Opportunities Found and Lost: Labor, Radicals, and the Early Civil Rights Movement," *Journal of American History* 75, no. 3 (1988): 786–811; Robert Rodgers Korstad, *Civil Rights Unionism: Tobacco Workers and the Struggle for Democracy in the Mid-Twentieth Century South* (Chapel Hill: University of North Carolina Press, 2003).

4. Van Gosse, "A Movement of Movements: The Definition and Periodization of the New Left," in *Companion to Post-1945 America*, ed. Jean-Christophe Agnew and Roy Rosenzweig (Malden, MA: Wiley and Sons, 2002), 277–302.

5. From the start, historians have been complicating the Montgomery-to-Memphis narrative. See, for example, William Chafe, *Civilities and Civil Rights: Greensboro, North Carolina and the Black Struggle for Freedom* (New York: Oxford University Press, 1981), a classic work that I failed to mention in my original article. As Kevin Boyle points out, by the early twenty-first century, an ever-growing body of literature made it possible to "solidify a different story." Kevin Boyle, "Redemption: Civil Rights, History, and the Promise of America," Hutchins Lecture, Center for the Study of the American South, University of North Carolina at Chapel Hill, January 18, 2011, https://vimeo.com/19129872.

6. For two of many examples, see Tomiko Brown-Nagin, *Courage to Dissent: Atlanta and the Long History of the Civil Rights Movement* (New York: Oxford University Press, 2011); and Margaret Burnham, "The Long Civil Rights Act and Criminal Justice," *Boston University Law Review* 95, no. 687 (2015): 687–712.

7. Margot Lee Shatterly, *Hidden Figures: The American Dream and the Untold Story of the Black Women Mathematicians Who Helped Win the Space Race* (New York: William Morrow, 2016).

8. Landon Storrs, *The Second Red Scare and the Unmaking of the New Deal Left* (Princeton, NJ: Princeton University Press, 2013); Jacquelyn Dowd Hall, *Sisters and Rebels: A Struggle for the Soul of the Modern South* (New York: W. W. Norton, 2019).

9. See, for example, William P. Jones, *The March on Washington: Jobs, Freedom, and the Forgotten History of Civil Rights* (New York: W. W. Norton, 2013), which argues that this 1963 rally for "*jobs* and freedom" was conceived by black trade unionists and left

feminists who had cut their political teeth in the 1940s. These men and women sought to link the new, direct action assault on segregation and disfranchisement with the ongoing struggle for access to jobs and union representation. For a local study that shows how the urban black workers who became grassroots activists in the southern movement in the 1960s had been primed for new forms of protest by decades of job-related struggles, see Max Krochmal, "An Unmistakably Working-Class Vision: Birmingham's Foot Soldiers and Their Civil Rights Movement," *Journal of Southern History* 76, no. 4 (2010): 923–960. For selections from a flourishing literature on black left feminism, see Dayo F. Gore, Jeanne Theoharis, and Komozi Woodard, eds., *Want to Start a Revolution? Radical Women in the Black Freedom Struggle* (New York: New York University Press, 2009).

10. Emilye Crosby, ed., *Civil Rights History from the Ground Up: Local Struggles, a National Movement* (Athens: University of Georgia Press, 2011), 1–39.

11. See, for example, Steven F. Lawson, "Long Origins of the Short Civil Rights Movement, 1954–1968," in *Freedom Rights: New Perspectives on the Civil Rights Movement*, ed. Danielle L. McGuire and John Dittmer (Lexington: University Press of Kentucky, 2011), 9–37.

12. Boyle, "Redemption."

13. Jeanne Theoharis, *A More Beautiful and Terrible History: The Uses and Misuses of Civil Rights History* (Boston: Beacon Press, 2018).

14. Hugh Davis, email message to author, August 9, 2018.

Introduction

Exploring NASA in the Long Civil Rights Movement

BRIAN C. ODOM

On July 15, 1969, the Reverend Ralph Abernathy and members of the Southern Christian Leadership Conference's Poor People's Campaign gathered outside the gates of the Kennedy Space Center in protest of the vast sums diverted from social programs here on Earth to the technological marvel now pointed at the Moon. To many Americans, the Apollo program represented the embodiment of the Enlightenment ideal of progress—an ideal grounded in the belief that all problems encountered in society were surmountable through the application of scientific principles. However, to the African American protestors standing alongside Abernathy and to many other critics across the country, Apollo represented a turning away from the plight of the poor and, with it, the abandonment of previous gains of the civil rights movement.

The intersection of the civil rights movement and the National Aeronautics and Space Administration (NASA) on that July day at Cape Canaveral, Florida, was neither the first nor the last. During the decade of the 1960s, NASA became something of a laboratory for social progress. Actions such as President John F. Kennedy's order for Equal Employment Opportunity in March 1961 and the creation of cooperative education programs between NASA and several Historically Black Colleges and Universities were representative of the concrete but limited engagement between the federal government and black graduates in technical disciplines. With the Apollo program, black graduates found the doors of economic opportunity gradually opening in many previously segregated occupations. White and black women also saw an increase in opportunities for themselves in the previously white, male-dominated fields of engineering and science. While the overall small increase in terms of numbers by the launch of Apollo 11 on July 16, 1969, left much to be desired, the pathways to new economic opportunity for black graduates and for women were enshrined in law and visible within the public discourse.

This collection of diverse essays is intended to promote a deeper, interdisciplinary exploration of the social history of NASA during the age of Apollo, developed around the ideas advocated by Jacquelyn Dowd Hall in her 2005 essay, "The Long Civil Rights Movement and the Political Uses of the Past." As a means of centralizing individual agency, Hall expanded the timeline of the civil rights movement and extended the scope to include the struggle for women's rights and labor. She also called upon historians to place their histories of the civil rights movement "in the context of a longer story," a move that would make that history "harder to simplify, appropriate, and contain."[1] Recent additions to space historiography and filmography also reflect the growing interest in racial and social aspects of space history. Using Hall's framework, the contributing authors seek to provide a deeper degree of analysis so that other stories might emerge from the richer context.

Applying this interdisciplinary approach to the experience of NASA during the Apollo program provided the authors an opportunity to recontextualize the experience of the space agency against the backdrop of social revolution, an economic transformation, and the growing Cold War with the Soviet Union. This approach reflects what former NASA chief historian Roger Launius described as the "New Aerospace History" and is representative of other works from the NASA History Office, including the *Societal Impact of Spaceflight*.[2] In his essay from that volume titled "Space History from the Bottom Up," Glen Asner echoes Hall's argument, noting that by "bringing ordinary people and social groups into our analysis," historians could "avoid reifying the concept of society and relegating masses of people to passive subjects of historical forces."[3]

Adjusting the view of the movement to the perspective of the Apollo program also offers new ways of thinking about both the impact of technology on society and the constraining environment offered in the Cold War context. The purpose of the following collection of essays is to address the role/relationship of NASA to the long civil rights movement, particularly, but not limited to, the Deep South (Huntsville, Alabama; Florida; Houston; Mississippi; and New Orleans) and to identify the impact of

NASA on the movement and the experience of those who were directly affected by the space program but also the impact of the movement on NASA's development in the period.

The chapters in part 1, "New Frameworks," address the larger conceptual themes of NASA's relationship to the civil rights movement in an effort to expand beyond the largely white-focused, hagiographical narratives of the space program. In the first chapter, Margaret Weitekamp explores the state of the "New Aerospace History" field. Here, Weitekamp offers a reevaluation of the state of the field in space history, looking especially at the influence of race, gender, and regional history. In response, Weitekamp suggests that attention should be paid to three major areas of growth in the field: individual and collective biography, fresh takes on technologies and cultural contexts, and international/global history. With the fiftieth anniversary of the first lunar landing in 2019, Weitekamp offers fresh thoughts on how historians should evaluate recent histories written since the last major reexamination in 2007, when it was occasioned by the fiftieth anniversary of the birth of the Space Age.

In chapter 2, P. J. Blount and David Molina trace NASA's attempted counternarrative of social value and a policy of liberal equality rooted in the concept of "all mankind." They consider whether this argument for NASA's value remains a salient one at present as the continued inequalities in American life are increasingly highlighted in the media, and as we face a historical moment in which activists and astronauts alike will be challenged to bridge the distance between Black Lives Matter and Mars. In chapter 3, Molina and Blount offer a contextualization of NASA's interlocutory role throughout the long civil rights movement by mobilizing these three themes to analyze a series of archival and cultural artifacts. The authors first analyze the rhetoric deployed by the Poor People Campaign's various mobilizations to show that the American space program was viewed with deep skepticism by the African American community and particularly within the context of ongoing struggles for black freedom. Second, they discuss the "distance" between the tropes of spatial disenfranchisement represented in the civil rights movement and the Moon missions to show how space exploration was portrayed as an acceleration of the marginalization of black spaces.

The three chapters that make up part 2, "Southern Context," explore the experience of the space agency in the Jim Crow South. Brenda Plummer examines the effect of the US space program on race relations in key areas of the South and the impact of that connection on popular culture. She also explores the intersection of the struggle for racial equality and aerospace exploration, as both constituted potent narratives of freedom in the American imaginary. Plummer disputes the assumption that NASA as an instrument of modernization was implicitly allied with the civil rights movement. While the transformation of parts of the Deep South undeniably broke up earlier political, economic, and cultural patterns, aerospace research and development helped inaugurate a successor regime that neither challenged the structural foundations of racial inequality nor guarded against the production of new disparities.

Matthew L. Downs explores the impact of Sun Belt–era federal development and the response of civic and commercial leaders to the civil rights movement, demonstrating how local leaders worked closely with government officials to attract and maintain such installations and the accompanying public and private investment. When federal officials and their representatives in Huntsville made clear that southern intransigence on civil rights would adversely affect the local, space-based economy, the city's civic leaders modulated their approach to civil rights in the hopes of ensuring continued support. Such action was particularly surprising given the overtly hostile response to the movement by Alabama's other local leaders and the state government. While Huntsville was not without conflict, the presence of the federal government, combined with the threat that southern resistance might lead to a withdrawal of federal support, led the city to a more moderate reaction when the city's local movement pressured for equality.

In chapter 6, Brian Odom surveys the implementation of equal employment opportunity at NASA's Marshall Space Flight Center in Huntsville. Odom contends that Marshall's strategy focused on recruiting qualified African American engineering students outside of Alabama and developing a partnership with the Association of Huntsville Area Contractors (AHAC) locally. By serving both as a catalyst for technical educational programs in the Huntsville community and as a clearinghouse for job opportunities and racial dialogue, AHAC facilitated a modicum of progress toward minority gains. During the civil rights movement, local activists such as Dr. Sonnie Hereford III and aerospace executives such as Brown Engineering Company's Milton K. Cummings brokered "backroom" agreements meant to improve Alabama's "image" problem while also expanding opportunities for black graduates.

Part 3, "International Context," explores NASA and the civil rights movement in an international, Cold War context. Cathleen Lewis argues that throughout the Cold War, race played an important role in foreign policy, with the United States painfully aware that its civil rights situation could have an adverse impact on foreign policy ambitions abroad. The USSR preyed on that US sensitivity, calling the country out on its failures. In the early 1980s, almost a decade after US foreign policy had all but abandoned race as a Cold War issue, the race issue reemerged, albeit briefly, when the USSR launched the first black man into space. The mission had two benefits for the USSR. First, it was retaliation against the United States for boycotting the 1980 Moscow Olympics. Second, it preempted the launch of the first American black astronaut. This final battle over race in the Cold War ultimately revealed American domestic progress and the hollowness of Soviet space stunts.

Keith Snedegar explores the impact of the civil rights movement on decisions related to NASA facilities outside the United States. Snedegar maintains that when Charles C. Diggs Jr., one of the founders of the Congressional Black Caucus, visited the NASA satellite tracking station at Hartebeesthoek, South Africa, in 1971, he discovered a racially segregated facility where technical jobs were reserved for white employees and black Africans essentially performed menial labor. Upon his return to the United States, the Detroit congressman embarked on a two-year struggle, first to improve workplace equity at the tracking station, and, later, for the closure of facility. NASA administration under James Fletcher was largely indifferent to demands for change at the station. It was only after Representative Charles Rangel proposed a reduction in NASA appropriations that the agency announced plans to end its working relationship with the white minority regime of South Africa. NASA's public statements suggested that a scientific rationale lay behind the station's eventual closure in 1975, but this episode clearly indicates that NASA was acting only under political pressure, and its management remained largely insensitive to global issues of racial equality.

The fourth set of essays address larger issues of race, gender, and the utilization of NASA technology on Earth. In chapter 9, Cyrus Mody argues that NASA and the Johnson Space Center experienced the 1970s through the paradox of "existential success." The Apollo program convinced other organizations that NASA engineers had "a competence which should be used" and therefore hired those engineers away and/or tapped NASA's expertise for their own organizational objectives. Meanwhile, the gap between Apollo and the shuttle meant NASA had reasons to accede to such demands, and few political resources to resist them. As a result, the possibility emerged, if briefly, for NASA to reorient its mission to the issues given currency by the civil rights movement: poverty (especially among ethnic minority communities), environmental justice, the urban dysfunctions created by white flight to the suburbs, and so forth. That that possibility soon disappeared, though, says much about the changing politics of race and civil rights in the 1980s and beyond.

Eric Fenrich studies the efforts of black activists and NASA to increase minority educational access that would lead to greater participation in the space program. According to Fenrich, the concurrence of the civil rights movement and the American space program reveals the two primary methods by which the advocates in the modern era have sought to advance the interests of African Americans. First, a negative project: the removal of formal barriers to the exercise of rights, more specifically, ending discriminatory practices in employment and education. Second, more positive efforts, such as equal employment opportunities and affirmative action, that place opportunities within the reach of historically disadvantaged people.

In chapter 11, Christina Roberts explores the perception that NASA performed poorly in hiring women during the long civil rights era, arguing that that perception is based on low recruitment numbers by comparison with many other federal agencies. Others blame NASA's poor recruitment efforts on an entrenched white male corporate culture that resisted hiring women and minorities into the early 1970s. While not denying NASA difficulties in the area, Roberts argues that what is missing from the historiography is a discussion about NASA's actual public outreach efforts to employ women scientists and engineers. Roberts contends that during the late 1950s to mid-1960s, NASA sought to transmit a message that women were welcome to apply and would attain professional science and engineering careers at NASA.

In his conclusion to the volume, Jonathan Coopersmith underscores the two major challenges of doing history—finding and preserving material, challenges particularly acute for subjects traditionally not collected by archives, such as minority movements. To prevent such future losses, Coopersmith explores how historians, archivists, and other stakeholders can encourage the creation and preservation of the widest possible range of appropriate records and histories, especially for historically underrepresented and underresearched areas and people in space exploration and exploitation.

These essays represent only a fraction of the work remaining in the expanded field of "New Aerospace History." Several glaring omissions include (but are not limited to) the intersection of NASA and the civil rights movement in other areas of the South, including Houston and the Kennedy Space Center; a comparative look at the impact in NASA's northern and western field centers; and a detailed examination of the impact on math and science education at black institutions of higher learning. The intention is that this volume will stimulate further interdisciplinary research into not only the social history of NASA but also the larger relationship between technology and American society.

Notes

1. Jacquelyn Dowd Hall, "The Long Civil Rights Movement and the Political Uses of the Past," *Journal of American History* 91 (March 2005): 1233–1263

2. Steven J. Dick and Roger D. Launius, eds., *Societal Impact of Spaceflight* (Washington, DC: NASA History Office, 2007).

3. Glen Asner, "Space History from the Bottom Up: Using Social History to Interpret the Societal Impact of Spaceflight," in Dick and Launius, *Societal Impact of Spaceflight*, 389.

PART I

NEW FRAMEWORKS

Space History Matures— and Reaches a Crossroads

MARGARET A. WEITEKAMP

In 2000, Roger D. Launius, then chief historian at the National Aeronautics and Space Administration (NASA), coined the phrase the "New Aerospace History" in a *Space Policy* article. In doing so, he distinguished that new body of scholarship from either the laudatory histories that became known as the Huntsville school or critiques of space flight written as histories. Rather, he argued, the New Aerospace History developed in the 1990s as a body of historical writing done by "professionally-trained scholars of differing ideologies and prerogatives who concentrate on questions other than whether or not space exploration is justifiable." His chosen name deliberately echoed the "New Social History," a body of American historical practice rooted in insights gleaned from the social movements of the 1960s and 1970s. Launius was well positioned to suggest that space history had attained a certain scholarly maturity. After serving as NASA's chief historian from 1990 to 2002, he took positions as a curator and later associate director at the Smithsonian National Air and Space Museum from 2002 to 2016. In those roles, he personally mentored many of the scholars whose work created the basis for the appellation.[1]

In 2006, I published my analysis of the New Aerospace History in the *Critical Issues in the History of Spaceflight* volume. In preparation for the conference that inspired the edited collection, Launius and his successor,

NASA chief historian Steven J. Dick, tasked me with evaluating the state of the field and its relationships to the various subdisciplines. Only a few years out of graduate school at the time, I was still freshly experienced in writing essays demonstrating the breadth and depth of my own reading and understanding. Indeed, such historiographical reflections have not only served as solid training for students learning the profession but also have been fundamentally characteristic of the academic field itself. As the Pulitzer Prize–winning historian Michael Kammen wrote in 1980 in his own assessment titled "The Historian's Vocation and the State of the Discipline in the United States," "Ever since . . . [the founding of the American Historical Association in 1884–85], the guild has had a penchant for introspection." Reviews of the state of the field provide the starting point for every research project and serve as periodic assessments of the rigor and vigor of the scholarly community as well.[2]

At the time, I concluded that "the insights of the New Social History have still been only incompletely incorporated into space history. This deficit is not attributable to a lack of source material, but rather to a limited perspective on what it would mean to integrate the story of race, class, ethnicity, and gender into space history more fully."[3] My resulting essay called on aerospace historians to engage the scholarly tools offered by gender studies, critical race studies, and other theoretically inflected humanities fields. Some people in the audience that day expressed skepticism about how useful that suggestion would be. Could fields so seemingly dependent on academic jargon offer much to empirical historians? A decade later, however, analytical frameworks that were first developed in the humanities have emerged from their ivory-tower origins into everyday language. Ordinary people in online communities regularly employ complex understandings of gender as a fluid category, as well as concepts of privilege, whiteness, and intersectionality. The deft use of such theoretically grounded scholarly tools is refreshingly common for emerging scholars at both the undergraduate and graduate levels. At the same time, aerospace historians have incorporated such concepts into compelling historical arguments built on close readings of documents and other archival sources. As the Cold War has receded further into history and different voices joined the conversation, scholars have also questioned the national focus that underlay many space history studies. Just as writers have revealed new historical actors ("hidden figures") in recent years, so too new research has widened the lens, revealing international contexts and exchanges overshadowed by Cold War preoccupations. As a result, a dozen years later I welcomed the opportunity to revisit and build upon my essay.

Recent work in space history builds on a strong analytical record. In the first decade of the twenty-first century, NASA's Headquarters History Office, helmed from 2003 to 2009 by Steven Dick, produced an impressive set of edited volumes exploring a broad set of scholarly questions. In addition to the "Critical Issues" conference and volume mentioned above, in September 2006 NASA's Headquarters History Office and the Smithsonian's National Air and Space Museum organized a meeting investigating the societal impact of spaceflight that convened in Washington, DC. Then in 2007, to mark the fiftieth anniversary of the first artificial satellite, the Soviet *Sputnik*, the same two organizations collaborated again, inviting "big picture" historians and scholars from "outside the usual circle of space history" to consider the Space Age's meaning "in the broadest sense of the word." Edited volumes documented the findings of both meetings for future scholars. An interdisciplinary volume, *Cosmos and Culture*, followed in 2009, exploring cosmic evolution from many different angles. Finally, in 2015 the NASA History Office organized a series of in-depth studies into an edited volume exploring the "relationship between science, technology, and society." Together, these substantial published volumes testify to the solid grounding in critical scholarly practice nurtured by that federal history office, now under the leadership of William "Bill" Barry.[4]

This retrospective essay was itself inspired by the innovative conference organized in March 2017 by the History Office of NASA's Marshall Space Flight Center. Inspired by Jacqueline Dowd Hall's pathbreaking essay, "The Long Civil Rights Movement and the Political Uses of the Past,"[5] participants considered the history of NASA programs and centers, including Marshall, as embedded in the history of the New South and the civil rights movement. Inspired by this creative scholarly linkage, the resulting conference drew together different communities of scholars and activists to consider in fresh ways how space history was embedded in larger social, cultural, political, and racial histories. The conversations that resulted modeled informed civil discourse, asking participants to respond thoughtfully to what they had learned. The new connections on display at that meeting got me thinking again about the assessment of the field that I had completed more than ten years earlier.

Examining the past dozen years or so of active scholarship, then, what has developed? What shapes this field now? In that compressed time frame, it can be difficult to discern broad patterns in categories of scholarship that most longer-term historiographies try to identify. Understanding that review essays cannot be comprehensive or exhaustive, however, I suggest here that attention should be paid to three major areas of growth in the field: individual and collective biography, fresh takes on technologies in their broader contexts, and international/global history. First, some of the most innovative work has been done in critical histories that document the stories of participants, whether through biography, collective biography, or social history. Second, there have recently been creative takes on traditional space history topics, including spacesuits, places of invention, and planetary exploration. Finally, new efforts to develop histories of international spaceflight have revealed how the field's Cold War mentality has long overshadowed global history frameworks. Notably, all three categories overlap significantly, as global or international history can no longer be meaningfully isolated as just one part of the field.

Biography and Social History

In the last ten years, the literary form of historical biography has emerged as a powerful way to make strong analytical arguments about the practice of science and technology, as well as the lives and influences of those who made spaceflight happen. Two sterling examples are James Hansen's exhaustive biography of the taciturn Apollo 11 moonwalker Neil Armstrong and Michael Neufeld's definitive biography of charismatic rocket engineer Wernher von Braun.[6] Each author relied on years of research to uncover the fullest possible histories of men with very public personas and complex personal lives. Hansen's investigation of Armstrong mined every possible documentary and living source—with the participation and blessing of the man himself, even as Armstrong continued his lifelong pattern of short, to-the-point answers without elaboration. The resulting story illustrates how one person took on the lasting responsibility of being the first man on the Moon. Neufeld's critical look at von Braun grappled with the complexities of his relationship with the Nazi party before and during World War II—as well as the implications for those associations in his later work and professional reputation. The result is not only a portrait of the man but also a clear-eyed investigation into how a complex history of remembering and forgetting became a key thread woven into the very fabric of the early US space program.

Recent biographies have attempted to contextualize and explain the impact of spaceflight professionals famous for their competence, such as Johnson Space Center director George Abbey (profiled by writer Michael Cassutt) and the first American woman in space, Sally K. Ride (written by journalist Lynn Sherr). Sherr's book explored the very private life of a public woman, whom she considered to be a personal friend but about whom she only learned certain personal details after the astronaut's untimely death in 2012. Her thoughtful and thoroughly researched account will be useful to historians. Andrew Jenks's analysis of Yuri Gagarin uses the myth and memory of the first man in space as a way to examine the cultural history of spaceflight in the Soviet Union. The historian Jared Buss's account of Willy Ley adds to the trend in space history of accomplishing nuanced historical writing through the form of biography.[7] To those accounts should be added memoirs and autobiographies. In particular, scholars should be grateful for *Rockets and People*, a four-volume translation of the memoirs of Russian space pioneer Boris Chertok. Translated by the man himself with guidance from series editor Asif Siddiqi, the resulting volumes were published by NASA in 2005, 2006, 2009, and 2011.[8] Such resources add depth to the materials available to future historians.

In addition to analytical books about single historical figures, strong new scholarship in the New Aerospace History also includes analyses of groups. *Inventing the Astronaut*, Matt Hersch's fresh take on the NASA astronaut corps as a labor force and a professional guild, should now be counted as essential reading for anyone doing human spaceflight history. In addition, scholars should take note of David Onkst's long-awaited dissertation on the engineers at Grumman Aerospace, whose labor building the Apollo lunar modules also became a story about the quick downturn in the aerospace labor market as Apollo wound down. Finally, in a collaborative effort about a collective group, Mike Neufeld's edited *Spacefarers* volume examined the social, cultural, and political significance of both astronauts and cosmonauts. Some of the strongest new work in space history, however, involves collective histories of previously unexamined groups.[9]

In my *Critical Issues* essay, I concluded that even given the influences of the New Social History on the New Aerospace History, "questions of race and ethnicity have been almost entirely ignored."[10] That is no longer the case. Excellent new books about race, culture, and spaceflight have shown that sources can be found to tell such stories in depth and with sensitivity. Radio-producer-turned-writer Richard Paul worked with his own oral histories and the historian Steven Moss's pathbreaking master's thesis to craft *We Could Not Fail*, a history of some of the first African American engineers working at NASA. (In full disclosure, I advised Paul's fellowship at the National Air and Space Museum.) In addition, Monique Laney's award-winning book *German Engineers in the Heart of Dixie* draws on oral history interviews and extensive documentary research to examine the integration of German immigrant rocket engineers into the southern town of Huntsville, Alabama. The resulting history examines race, religion, culture, politics, and technology along with the history of the civil rights movement.[11]

Thanks to the success of a best-selling book and a major motion picture of the same name, a shorthand now exists for little-known historical women in aerospace: "hidden figures." In that book, Margot Lee Shetterly documented women who have been well known in the historical community but not previously identified as significant in their own right. Several similar works have added parallel stories to the literature. George D. Morgan's personal reflections on his mother's career, *Rocket Girl*, mixed space history with her individual history refining rocket fuels at North American Aviation. Nathalia Holt's *Rise of the Rocket Girls* documented the women computers at the Jet Propulsion Laboratory, while Dava Sobel's *The Glass Universe* recounted the history of women assistants at the Harvard College Observatory. An ongoing project continues to examine the notes and other evidence of the women working at the Harvard Observatory. As scholars continue to investigate questions of race and gender, new histories will surely be added to this category.[12]

Technologies in Their Broader Contexts

The second major area of new work in space history is really an accumulation of innovative takes on traditional topics. These are not the only topics that have been well developed in space history, but they offer some telling examples. In some cases, fresh interpretations of spaceflight subjects have come from interdisciplinary work. Architect Nicholas de Monchaux's thoughtful book *Spacesuit* drew its form from the twenty-one layers of the Apollo lunar spacesuit, weaving together twenty-one separate stories inspired by that piece of wearable technology and architecture. Before she retired from the Smithsonian Institution, Mandy Young also completed her authoritative book about the spacesuits in the National Air and Space Museum's collection, illustrated with photographs by Mark Avino. David Mindell looked at a different spaceflight technology: the Apollo guidance computer. His interpretations of the intersections between the human beings flying in Apollo spacecraft and the computers being used to guide them goes beyond the usual history of how the things were built and used to consider the broader significance of these spaceflight technology innovations.[13]

Scholars have also begun to reexamine how the space program fits into cultural history. Kendrick Oliver's *To Touch the Face of God* explores the many historical connections between NASA, the divine, and religion. In contrast, Matt Tribbe's *No Requiem for the Space Age* argues that the end of the Apollo lunar landing program coincided with a cultural decline in faith of a very different kind: belief and trust in science. This, he suggests, explained the turn away from such big science and technology programs amid the flowering of the counterculture. Neil Maher's *Apollo in the Age of Aquarius* sees less inherent conflict between the space program and its cultural context. Maher works instead to reintegrate the space program into the broader social and cultural history of the late 1960s and 1970s. Teasel Muir-Harmony draws on this scholarship as well as the collections of the Smithsonian Institution in her reflections on the fiftieth anniversary of the Apollo program as told through fifty artifacts in the national collection. By including not only spacecraft and personal equipment but also the chairs used in the televised Kennedy-Nixon presidential debate in 1960 or a collection can used by the Southern Christian Leadership Conference (SCLC) later that same decade, the lens widens, capturing the history of spaceflight efforts as a part of the long 1960s.[14]

Creative new scholarship has also been done in the areas of visual and material culture. Elizabeth Kessler's *Picturing the Cosmos* considers the images produced by the Hubble Space Telescope not only as scientific findings but also as aesthetic images, which have become immensely popular with the public. Megan Prelinger's *Another Science Fiction* tapped the extraordinary visual collections of advertisements held by the Prelinger Library to look at advertising, especially as aimed at engineers, in the early years of the Space Age. Scholars have also taken new looks at how the physical artifacts of spaceflight have been displayed, discarded, and abandoned, both on Earth and in space. In a detailed look at Apollo program memorabilia, the two authors of *Marketing the Moon* used their personal collections of collectibles to tell the story of how the Apollo program was sold to the public—and then, in turn, used by advertisers to tap into the zeitgeist of the late 1960s. We can also look forward to more new scholarship in the fields of visual culture, material culture, and built environments.[15]

This present volume, *NASA and the Long Civil Rights Movement*, examines the local histories of NASA centers as intersecting with the civil rights histories of Alabama and the American South. In the last ten years, innovative work has also been done in examining the sites of aerospace and technological development as places of invention. Paul Ceruzzi, a historian of computing, focused on Tyson's Corner, Virginia, as a place that transformed from a sleepy farm town into the crossroads of the Internet and a hub of computing. In 2009, Joe Bassi completed a dissertation analyzing how Boulder, Colorado, became a center for space and atmospheric science. Finally, at the University of Southern California, in association with the Huntington Library, Peter Westwick led a study of the aerospace industry in Southern California. The work resulted in an exhibit and an accompanying edited volume, both called *Blue Sky Metropolis: The Aerospace Century in Southern California*.[16]

In the last ten years, the end of the space shuttle program has also offered scholars an opportunity to begin to account for its full history. The historian T. A. Heppenheimer completed his two-volume prehistory of the space shuttle, focusing on the initial technology decision and system development, ending in 1981 with the vehicle's first flights. In addition, former shuttle engineer Dennis Jenkins has written a multivolume history of the actual missions flown by the reusable Space Transportation System. For future scholars, these will be invaluable and authoritative resources. Valerie Neal has also written two books about the shuttle. The first expanded upon each of space shuttle *Discovery's* individual missions; the second offers a larger analytical history about how the overall program was sold, explained, and rationalized. For his part, Jim David has investigated the relationships that NASA had with the defense and clandestine agencies in the United States during the space shuttle period. One looks forward to additional scholars bringing a critical eye to this period.[17]

The maturity of space history as a scholarly field can also be seen in the ways that planetary exploration has been analyzed. Contributions such as Peter Westwick's history of the Jet Propulsion Laboratory and Eric Conway's analysis of the laboratory's Mars missions reveal the bureaucratic corridors that exploratory missions must successfully traverse before they ever leave the Earth—and how institutions foster and support science missions. Likewise, Michael Neufeld's analysis of the long prelaunch journey of the *New Horizons* probe, even before it traveled all the way to Pluto and beyond, is ably supported by the first-person account published by *New Horizons* principal investigator Alan Stern in collaboration with writer David Grinspoon. Work in the history of planetary science is moving beyond the simple recording of missions to ask bigger questions about the complex interplay of institutions, budgets, people, design, science, and technology.[18]

International and Global History

The last ten years have seen a broad turn in the practice of American history to considering the United States in the world. *NASA in the World*, cowritten by John Krige, Angelina Long Callahan, and Ashok Maharaj, offers a fine example of that lens applied to space history. The well-researched and detailed volume offers an extensive history of NASA collaborations with Western Europe, as well as with the Soviet Union, Japan, and India. In the concluding section, Krige turns attention to both the international cooperation undergirding the International Space Station and the guidance intended to forestall problematic technology transfers that is the International Traffic in Arms Regulations (ITAR). Such restrictions originated as Cold War provisions, which were later adapted to the modern war on terror. The overall study illuminates the long history of NASA's international agreements and projects. In-depth studies of the Chinese human spaceflight program, the Indian space program, or other spaceflight efforts around the world will be welcomed as they are produced.[19]

Some of the strongest scholarship in the field of space history in recent years has been done about the cultural history of spaceflight in the Soviet Union/Russia. As recently as the 1990s, Asif Siddiqi was working to compile the names, dates, and events that comprised the first comprehensive history of the Soviet space program. Now a number of scholars, himself included, have deepened that knowledge through biography and cultural history.[20]

Moving away from the point of view of either of the Cold War superpowers, however, Alexander C. T. Geppert's group at the Free University of Berlin has inspired important work in space history from a European perspective. Beginning in 2010 and until 2016, Geppert directed the Emmy Noether Research Group called "The Future in the Stars: European Astroculture and Extraterrestrial Life in the Twentieth Century" at Freie Universität Berlin. Geppert's conferences and volumes have contributed substantially to developing the field of European space history.[21] In particular, much of this work has centered around the idea of "astroculture," a shorthand for the cultural history of spaceflight, broadly considered.

In another important development, the publication of Israeli author Deganit Paikowsky's *The Power of the Space Club* offers a historically grounded analytical perspective for framing current international space policy discussions. Paikowsky used the case studies of various international space efforts to analyze why having an internal spaceflight capacity has been seen by nation-states as an important marker of their prestige and status on the world stage. As more and more nations join the space club (Bangladesh launched its first communications satellite in 2018), her study provides a useful tool for policy makers as well as a must-read for future scholars considering national and international space programs.[22]

Scholars have already begun answering the appeal made in Asif Siddiqi's 2010 essay in *Technology and Culture* calling for a new global history of spaceflight. Siddiqi, writing from his perspective as both an American-trained scholar and one of the foremost historians of the Soviet Union/Russian space program, stepped back in 2010 to publish a directive for space history as a whole. Arguing that "our understanding of the half-century of space travel is still firmly rooted in the framework of the national imagination," he suggests that "the maturation of other national space programs—those of China, Japan, and India, for example—will require us to approach space history with new lenses as more and more 'new' narratives join the old cold-war-centered approach to space history." Siddiqi points out that international transfers of people, knowledge, and technologies are not new (think, for instance, of the postwar integration of German vehicles and engineers who helped to develop rocketry in the United States). But past scholars have told these stories as part of nationally focused histories, without considering them as global influences. In contrast, Siddiqi calls on historians "to incorporate a broader matrix of approaches, including, particularly, the highlighting of global flows of actors and knowledge across borders, communities, and identities . . . [resulting in for] the first time a global and transnational history of rocketry and space travel." By "decentering" the focus, Siddiqi argues, "one might expect a multitude of smaller, local, and ambiguous processes and meanings to become visible."[23]

Indeed, numerous scholars are already writing histories on those terms, driven by an inherent interest in globalization and its uneven effects. For instance, Martin Collins's history of the Iridium satellite, Motorola, and satellite phones explicitly grounds its analysis in the complicated construction and implications of "the global." Likewise, Paul Ceruzzi's new concise history of GPS grapples with the worldwide implications of this now ubiquitous space-based technology and its analogous systems in other countries. And Michael Neufeld's brief volume *Spaceflight*, part of MIT Press's "Essential Knowledge" series, deliberately integrates the newest scholarship on space programs outside of the United States and Russia into a story that used to be told primarily in reference to those two superpowers.[24]

The Current Moment

As historians look to the future, one finds the aerospace industry in the United States both thriving and at a crossroads. Many historians of technology are already watching the developing field of commercial human spaceflight. In 2004, Scaled Composites Tier One won the $10 million Ansari X Prize for the two successful flights of SpaceShipOne, the first privately developed suborbital human spaceflight vehicle. Commercially available space tourism seemed to be imminent until SpaceShipTwo crashed during a test flight in 2014, killing one of the two pilots on board. In the summer of 2017, the characteristically frank designer of SpaceShipOne, Burt Rutan, openly expressed his frustration with the failure of the space tourism business to live up to its promised schedules. At the same time, the next-generation astronomical instrument, the James Webb Space Telescope, has been delayed in its launch even as the Hubble Space Telescope has been operating on reduced gyroscopic power and the Kepler planet finder went offline in November 2018 after running out of fuel. The next Mars lander, InSight, landed and began examining the interior structure of Mars that same month.[25]

In the last decade, other businesses have pursued different aspects of spaceflight. In May 2016, NASA pressurized the Bigelow Expandable Activity Module (BEAM), an inflatable test module developed from old NASA plans using new NASA money. And in what has been called a new space race, Internet billionaires have poured resources into developing new launch vehicles. While Jeff Bezos's Blue Origin has made quietly steady progress, Elon Musk's SpaceX has seemingly perfected the recovery of spent first stages of Falcon 9 launch vehicles via controlled upright landings on robotic seafaring barges. At the same time that SpaceX is developing the Dragon spacecraft for cargo and crew supplies to the International Space Station, Boeing is also developing a next-generation human spaceflight vehicle, the CST-100 Starliner. Whether these efforts represent true disruptions in techniques, technologies, or business models depends upon who is asked. As more time passes, however, this new public face of commercial spaceflight will continue to evolve from a topic of contemporary space policy to being the subject of historical consideration. As it does, Launius's recent studies of historical analogs for commercial space, comparing these new business ventures to the development of federally subsidized railroads and airlines, provide useful guides.[26]

At the same time, historians will be watching the next direction for national and international space efforts. In 2017, President Donald Trump issued an executive order reconstituting the National Space Council as a means of guiding national efforts in three key areas: civil space, commercial space, and national security space. The policy decisions that emerge from that executive body's deliberations will guide space policy—and the vast existing businesses already servicing this sector—in the United States in the near future. And in keeping with Siddiqi's call for a global history of spaceflight, historians should also be paying attention to international spaceflight efforts—and the true global nature of seemingly national ones. Specifically, historians will be watching how international alliances weather the end of the International Space Station sometime in the next decade. More so, the historical roots of spaceflight efforts in nations around the world remain rich areas for scholarly exploration.

Conclusion

I end this historiographic assessment with a note that is somewhat discouraging for this writer but hopeful for the overall field: this essay will be obsolete as soon as it appears. Many prominent aerospace authors have been working on significant pieces aimed to coincide with the fiftieth anniversary of Apollo 11's first landing of humans on the Moon in July 1969. As a result, however, whoever writes the next version of this essay should begin with contributions from 2019, which promises to be an *annus mirabilis* for space history.

More important, fiftieth anniversaries mark a significant shift, from memory to history. It is a sad reality that fewer of the people who made the history of the Apollo lunar landings in the 1960s are around to celebrate the fiftieth anniversary than were present at the fortieth anniversary celebrations in 2009. Likewise, more and more of the scholars who address space history topics in the future will only know those events as history, not lived memory. There is potential danger is this transition. Just after the fiftieth anniversary of the end of the American Civil War, for instance, various organizations began to enshrine the myth of the Southern "Lost Cause" in statues and memorials on battlefields throughout the theater of war. Those distortions of history remain politically contested almost a hundred years later. In the fifty years since the Moon landings, for example, conspiracy theories about the lunar landings and even the idea that the world is flat have proliferated. For space history, there is distinct value lost as those who participated in or otherwise witnessed the golden age of human spaceflight retire. Yet the passage of the fiftieth anniversary also has great potential for growth in rigorous scholarship. A new generation of scholars will be able to analyze the history of spaceflight with fresh eyes.

Several initiatives are in the works to foster this. In anticipation of the Apollo 11 anniversary, the "To Boldly Preserve" conference gathered historians, librarians, and archivists in 2017 to consider how best to manage and preserve the physical and digital history of the Space Age. That initiative is in the process of becoming a more permanent collaboration that will support space history endeavors. At the same time, the NASA History Office and the National Air and Space Museum's Space History Department marked the Apollo 11 fiftieth anniversary year by once again collaborating to convene a gathering of space history scholars, this time for a workshop exploring the future of the field. The brainstorming session aimed to inspire a new generation of scholars to mine the untapped potential of space history topics in future dissertations, articles, and books. Yet all of that material will emerge after this chapter appears. That bodes well for a mature historical field—as well as for the potential for rigorous scholarship to reach a wider reading audience who might be inspired by the anniversary to investigate space topics.

Notes

1. Roger D. Launius, "The Historical Dimension of Space Exploration: Reflections and Possibilities," *Space Policy* 16 (2000): 23–38; Eric Foner, ed., *The New American History* (Philadelphia: Temple University Press, 1990).

2. Michael Kammen, ed., *The Past Before Us: Contemporary Historical Writing in the United States* (Ithaca, NY: Cornell University Press, 1980), 19.

3. Margaret A. Weitekamp, "Critical Theory as a Toolbox: Suggestions for Space History's Relationship to the History Subdisciplines," in *Critical Issues in the History of Spaceflight*, ed. Steven J. Dick and Roger E. Launius (Washington, DC: NASA Office of External Relations, 2006), 549–572.

4. Dick and Launius, *Critical Issues in the History of Spaceflight*; Steven J. Dick and Roger D. Launius, eds., *Societal Impact of Spaceflight* (Washington, DC: NASA Office of External Relations, 2007); Steven J. Dick, *Remembering the Space Age* (Washington, DC: NASA Office of External Relations, 2008), x; Steven J. Dick and Mark L. Lupisella, eds., *Cosmos & Culture: Cultural Evolution in a Cosmic Context* (Washington, DC: NASA, 2009); Steven J. Dick, ed., *Historical Studies in the Societal Impact of Spaceflight* (Washington, DC: NASA, 2015), vii.

5. Jacqueline Dowd Hall, "The Long Civil Rights Movement and the Political Uses of the Past," *Journal of American History* 91 (March 2005): 1233–1263.

6. James R. Hansen, *First Man: The Life of Neil A. Armstrong* (New York: Simon & Schuster, 2005); Michael J. Neufeld, *Von Braun: Dreamer of Space, Engineer of War* (New York: Knopf, 2007).

7. Michael Cassutt, *The Astronaut Maker: How One Mysterious Engineer Ran Human Spaceflight for a Generation* (Chicago: Chicago Review Press, 2018); Jared S. Buss, *Willy Ley: Prophet of the Space Age* (Gainesville: University Press of Florida, 2017); Lynn Sherr, *Sally Ride: America's First Woman in Space* (New York: Simon & Schuster, 2015); Andrew L. Jenks, *The Cosmonaut Who Wouldn't Stop Smiling: The Life and Legend of Yuri Gagarin* (DeKalb: Northern Illinois University Scholarly Press, 2012).

8. Boris Chertok, *Rockets and People*, vol. 1 (Washington, DC: NASA, 2005); Boris Chertok, *Rockets and People*, vol. 2, *Creating a Rocket Industry* (Washington, DC: NASA, 2006); Boris Chertok, *Rockets and People*, vol. 3, *Hot Days of the Cold War* (Washington, DC: NASA, 2009); Boris Chertok, *Rockets and People*, vol. 4, *The Moon Race* (Washington, DC: NASA, 2011).

9. Matthew Hersch, *Inventing the American Astronaut*, Palgrave Studies in the History of Science and Technology (London: Palgrave Macmillan, 2012); David Onkst, "The Triumph and Decline of the 'Squares': Grumman Aerospace Engineers and Production Workers in the Apollo Era, 1957–1973" (PhD diss., American University, 2011); Michael J. Neufeld, ed., *Spacefarers: Images of Astronauts and Cosmonauts in the Heroic Era of Spaceflight* (Washington, DC: Smithsonian Institution Scholarly Press, 2013). In the interest of full disclosure, one of my own essays appears in the *Spacefarers* volume.

10. Weitekamp, "Critical Theory as a Toolbox," 564.

11. Richard Paul and Steven Moss, *We Could Not Fail: The First African Americans in the Space Program* (Austin: University of Texas Press, 2015); Monique Laney, *German Rocketeers in the Heart of Dixie: Making Sense of the Nazi Past during the Civil Rights Era* (New Haven, CT: Yale University Press, 2015).

12. Margot Lee Shetterly, *Hidden Figures: The American Dream and the Untold Story of the Black Women Who Helped Win the Space Race* (New York: William Morrow, 2016); George D. Morgan, *Rocket Girl: The Story of Mary Sherman Morgan, America's First Female Rocket Scientist* (Westminster, MD: Prometheus Books, 2013); Nathalia Holt, *Rise of the Rocket Girls: The Women Who Propelled Us, from Missiles to the Moon to Mars* (New York: Little, Brown, 2016); Dava Sobel, *The Glass Universe: How the Ladies of the Harvard Observatory Took the Measure of the Stars* (New York: Viking, 2016); Alex Newman, "A Team of Women Is Unearthing the Forgotten Legacy of Harvard's Women 'Computers,'" *The World*, Public Radio International, July 27, 2017, https://www.pri.org/stories/2017-07-27/team-women-are-unearthing-forgotten-legacy-harvard-s-women-computers.

13. Nicholas de Monchaux, *Spacesuit: Fashioning Apollo* (Cambridge, MA: MIT Press, 2011); Amanda Young, *Spacesuits: Within the Collections of the Smithsonian National Air and Space Museum* (Brooklyn, NY: PowerHouse Books, 2009); David Mindell, *Digital Apollo: Human and Machine in Spaceflight* (Cambridge, MA: MIT Press, 2011).

14. Kendrick Oliver, *To Touch the Face of God: The Sacred, the Profane, and the American Space Program, 1957–1975*, New Series in NASA History (Baltimore: Johns Hopkins University Press, 2012); Matt Tribbe, *No Requiem for the Space Age: The Apollo Moon Landings and American Culture* (Oxford: Oxford University Press, 2014); Neil M. Maher, *Apollo in the Age of Aquarius* (Cambridge, MA: Harvard University Press, 2017); Teasel Muir-Harmony, *Apollo to the Moon: A History in 50 Objects* (Washington, DC: National Geographic, 2018).

15. Elizabeth S. Kessler, *Picturing the Cosmos: Hubble Space Telescope Images and the Astronomical Sublime* (Minneapolis: University of Minnesota Press, 2012); Megan Prelinger, *Another Science Fiction: Advertising the Space Race 1957–1962* (New York: Blast Books, 2010); David DeVorkin and Michael Neufeld, "Space Artifact or Nazi Weapon?: Displaying the Smithsonian's V-2 missile, 1976–2011," *Endeavour* 35 (December 4, 2011): 187–195; Roger D. Launius, "Abandoned in Place: Interpreting the U.S. Material Culture of the Moon Race," *Public Historian* 31 (August 2009): 9–38; Alice Gorman, "The Archaeology of Orbital Space," *Australian Space Science Conference 2005* (Melbourne: RMIT University, 2005), 338–357; David Meerman Scott and Richard Jurek, *Marketing the Moon: The Selling of the Apollo Lunar Program* (Cambridge, MA: MIT Press, 2014); Jennifer Levasseur, "Pictures by Proxy: Images of Exploration and the First Decade of

Astronaut Photography at NASA" (PhD diss., George Mason University, 2014); Lisa Ruth Rand, "Orbital Decay: Space Junk and the Environmental History of Earth's Planetary Borderlands" (PhD diss., University of Pennsylvania, 2016); Layne Karafantis, "Under Control: Constructing the Nerve Centers of the Cold War" (PhD diss., Johns Hopkins University, 2016).

16. Paul Ceruzzi, *Internet Alley: High Technology in Tyson's Corner, 1945–2005*, Lemelson Center Studies in Invention and Innovation Series (Cambridge, MA: MIT Press, 2008); Joseph Bassi, "Creating a Scientific Peak: How Boulder, Colorado, Became a World Center for Space and Atmospheric Science, 1945–1965" (PhD diss., University of California, Santa Barbara, 2009); Peter J. Westwick, ed., *Blue Sky Metropolis: The Aerospace Century in Southern California*, Western Histories (San Marino, CA: Huntington Library and University of California Press, 2012).

17. T. A. Heppenheimer, *History of the Space Shuttle*, vol. 2, *Development of the Space Shuttle, 1972–1981* (Washington, DC: Smithsonian Books, 2010); Dennis R. Jenkins, *Space Shuttle: Developing an Icon 1972–2013* (Forest Lake, MN: Specialty Press, 2017); Valerie Neal, *Discovery: Champion of the Space Shuttle Fleet*, Smithsonian Series (Minneapolis: Zenith Press, 2014); Valerie Neal, *Spaceflight in the Shuttle Era and Beyond: Redefining Humanity's Purpose in Space* (New Haven, CT: Yale University Press, 2017); James E. David, *Spies and Shuttles: NASA's Secret Relationships with the DoD and CIA* (Gainesville: University Press of Florida, 2015).

18. Peter Westwick, *Into the Black: JPL and the American Space Program, 1976–2004* (New Haven, CT: Yale University Press, 2006); Erik M. Conway, *Exploration and Engineering: The Jet Propulsion Laboratory and the Quest for Mars*, New Series in NASA History (Baltimore: Johns Hopkins University Press, 2015); Michael J. Neufeld, "First Mission to Pluto: Policy, Politics, Science, and Technology in the Origins of New Horizons, 1989–2003," *Historical Studies in the Natural Sciences* 44, no. 3 (2102): 234–276; Alan Stern and David Grinspoon, *Chasing New Horizons: Inside the Epic First Mission to Pluto* (New York: Picador, 2018). See also Roger D. Launius, ed., *Exploring the Solar System: The History and Science of Space Exploration* (New York: Palgrave Macmillan, 2013).

19. John Krige, Angelina Long Callahan, and Ashok Maharaj, *NASA in the World: Fifty Years of International Collaboration in Space*, Palgrave Studies in the History of Science and Technology (London: Palgrave Macmillan, 2013).

20. Asif A. Siddiqi, *Challenge to Apollo: The Soviet Union and the Space Race, 1945–1974* (Washington, DC: NASA History Division, 2000); Asif A. Siddiqi, *The Red Rockets' Glare: Spaceflight and the Russian Imagination, 1857–1957*, Cambridge Centennial of Flight (Cambridge: Cambridge University Press, 2010); Slava Gerovitch, *Soviet Space Mythologies: Public Images, Private Memories, and the Making of Cultural Identity* (Pittsburgh: University of Pittsburgh Press, 2015).

21. Alexander C. T. Geppert, ed., *Imagining Outer Space: European Astroculture in the Twentieth Century*, Palgrave Studies in the History of Science and Technology (London: Palgrave Macmillan, 2012); Alexander C. T. Geppert, ed., *Limiting Outer Space: Astroculture after Apollo*, Palgrave Studies in the History of Science and Technology (London:

Palgrave Macmillan, 2018). A future volume will complete the trilogy: *Militarizing Outer Space: Astroculture and Dystopia in the Cold War*.

22. Deganit Paikowsky, *The Power of the Space Club* (Cambridge: University of Cambridge Press, 2017).

23. Asif A. Siddiqi, "Competing Technologies, National(ist) Narratives, and Universal Claims: Toward a Global History of Space Exploration," *Technology and Culture* 51 (April 2010): 425–443, quotes from 425 and 426.

24. Martin Collins, *A Telephone for the World: Iridium, Motorola, and the Making of a Global Age* (Baltimore: Johns Hopkins University Press, 2018); Paul Ceruzzi, *GPS*, MIT Press Essential Knowledge Series (Cambridge, MA: MIT Press, 2018); Michael J. Neufeld, *Spaceflight*, MIT Press Essential Knowledge Series (Cambridge, MA: MIT Press, 2018).

25. Jeff Foust, "'What the Hell Happened?': The Rise and Fall of Suborbital Space Tourism Companies," *SpaceNews*, June 5, 2017, http://spacenews.com/what-the-hell-happened-the-rise-and-fall-of-suborbital-space-tourism-companies/.

26. Roger D. Launius, "The Railroads and the Space Program Revisited: Historical Analogues and the Stimulation of Commercial Space Operations," *Astropolitics: The International Journal of Space Politics & Policy* 12 (2014): 167–179; Roger D. Launius, *Historical Analogs for the Stimulation of Space Commerce*, Monographs in Aerospace History, no. 54 (Washington, DC: NASA, 2014).

Bringing Mankind to the Moon

The Human Rights Narrative in the Space Age

P. J. BLOUNT AND DAVID MIGUEL MOLINA

The single most cliché phrase in space exploration is likely Neil Armstrong's words as he stepped from Apollo 11's lunar module to the surface of the Moon: "One small step for man, one giant leap for mankind." These words, a scripted apex to the classical Space Age bookended by the orbiting of *Sputnik I* and the Moon landing that placed Armstrong on the lunar surface, have been replayed, remixed, co-opted, copied, and satirized, and have taken on a variety of meanings through the global zeitgeist. Despite the constant reproduction of meaning, the original sentiment contains a fascinating notion about the distinction between man and mankind, and the representational nature of Armstrong, the explorer, as an emissary for the human population.

Armstrong's words were chosen to imbue a specific political and social content not just to his step onto the lunar terrain or to the Moon landing itself but also to the entire US space enterprise. The phrase contained a deep ideological terrain intended to link humankind to the US space project, thereby linking US values to the imagined destiny of the global population. The vision of Armstrong stepping onto the Moon, beamed globally through television, was the image of US ideological dominance in the Cold War, and Armstrong's words were a statement of the universality of that ideology.

In this context, the use of the word "mankind" is not a coincidence. Indeed, the term has a specific pedigree within the rhetoric of domestic and international space activities. Its use by Armstrong was meant to contextualize the United States' Moon landing into an international historical narrative framing of space activities as "for the benefit of all mankind," and to fill that framing with American liberal ideology as opposed to Soviet socialist ideology.

This chapter, the first of two complementary chapters, is primarily concerned with the US project of framing space exploration activities as for the "benefit of all mankind." This framing established a supranational ethic rooted in the emergent discourse of universal human rights, which the United States then domesticated as emblematic of its Western liberal value system. The first section of this chapter traces the Cold War origins of the phrase "for the benefit of all mankind" in the aftermath of *Sputnik I* as it evolved from a presidential address, to federal legislation, and to international law. This section argues that the link between space activities and human rights was primarily concerned with increasing national and international security. Next, the chapter turns its attention to the Moon's *Mare Tranquillitatis*, where the United States implanted this concept on the lunar surface in order reify a supranational dedication to global humanistic space activities and, simultaneously, to advance US values and ideology through a domesticated version of the principle. The final section of the chapter posits the idea of a "long space age" as a way to break away from the ideological content held between *Sputnik I* and Apollo 11. This section specifically looks at the evolution of the US space program and its evolving commitment to human and civil rights. The 1957 orbiting of *Sputnik I* was a profound event that shocked the population of the Earth. It provided an epochal vantage point for simultaneously observing the population's globality while confirming its vulnerability.[1] As the world looked to the sky and listened to the beeping of *Sputnik*, there was recognition that for the first time in centuries, a new, unventured territory had been opened. At the same time, *Sputnik* demonstrated the necessary precursor for intercontinental ballistic missiles and a human ability to effectively extend the annihilating threat of nuclear war to global proportions. *Sputnik*'s sublime beep was a message of equal parts hope and fear.

In the face of this breakthrough in Soviet technology, the United States' immediate response to *Sputnik* was to reinforce its initial framing of the

US space program as a peaceful, scientific, civilian enterprise. Just two years prior, as both the United States and the USSR committed to place an artificial satellite in orbit as part of the International Geophysical Year (IGY), the United States dedicated its nascent space exploration program to open, scientific, and peaceful exploration.[2] Thereafter, the United States consistently insisted on the civil nature of its space program in contrast to the USSR's explicitly military venture, even to the detriment of losing the so-called Space Race.[3] This strategy sought to salvage diplomatic gains from a military lag: as the United States fell farther behind the USSR in rocket development, the United States increasingly sought to push for a heightened notion of space exploration for peaceful, scientific benefits.[4] As a result, the progress of humankind as such—expressed through the term "mankind"—became implicated into the space exploration project as a grounding ethos for articulating the basic motives of the Space Age.

The ascription of Cold War technology to global benefit and civil purpose—as opposed to national and military purposes—was not a new discursive strategy for US policy makers. The explicit foregrounding of mankind's stake in the benefits of US military technology can be proximally traced to earlier attempts to reframe nuclear proliferation for post–World War II audiences at home and abroad. In his 1953 "Atoms for Peace" speech to the United Nations General Assembly (UNGA), President Eisenhower stated, "the United States knows that if the fearful trend of atomic military buildup can be reversed, this greatest of destructive forces can be developed into a great boon, for the benefit of all mankind."[5] This would mark the first time the phrase "for the benefit of all mankind" would explicitly appear in US presidential discourse.[6] Eisenhower would renew this call in October 1957 just weeks before the launch of *Sputnik I*.[7] The following spring the phrase "for the benefit of all mankind" entered into space exploration discourse via Eisenhower's statements accompanying the Science Advisory Committee's "Introduction to Outer Space."[8] In the 1960 State of the Union address, Eisenhower fixed the peaceful use and common benefit of space in the imaginary of a national public, expressly stating that space sciences are "important foundation-stones for more extensive exploration of outer space for the ultimate benefit of all mankind."[9]

Interestingly, Eisenhower only uses "for the benefit of all mankind" one other time between his October 1957 address on atomic energy and the March 1958 Science Advisory Committee report. Crucially, this other instance—situated in the phrase's pivot from nuclear to Space Age discourse—pertains to neither, yet indexes a common link between Cold War security and the contested terrain of human rights. In December 1957, Eisenhower released Proclamation 3213, which declared December 10, 1957, to be UN Human Rights Day.[10] In this proclamation, he calls on citizens to publicly read both the Universal Declaration of Human Rights and the US Bill of Rights, and affirms that the United States is "dedicated to the principle of freedom," and that, as such, the nation should "draw strength from our own experience of liberty to use our new resources for the benefit of all mankind."[11] This statement connects the supranational concept of human rights that emerged after World War II to the uniquely American core of civil rights embedded in US constitutional order as a way to amplify American notions of human rights as a universalizing mechanism for liberal democratic progress at the supranational level.

The National Aeronautics and Space Act of 1958 (the Space Act) formally adopted this supranational ethic into US law by making it "the policy of the United States that activities in space should be devoted to peaceful purposes for the benefit of all mankind."[12] There are two important things happening in this policy statement. The first is that the United States is attempting to frame its own space activity as peaceful by drawing on international themes flowing from the UN Charter. At the same time, the phrase "for the benefit of all mankind" becomes domesticated as a narrative of US liberal progress gone global—that is, as a national commitment to "all mankind" recast in the celestial image of *homo americanus*.

After having adopted the "benefits" language as domestic policy, the United States then began to reroute this framing of space activities as an international norm. To this end, President Eisenhower, a Republican, asked then senator Lyndon Johnson, a Democrat, "to go to the United Nations and make a speech which showed that the Executive and Legislative branches were unified in promoting space for peaceful purposes, for the benefit of mankind."[13] This unified approach was an attempt to get the Soviet Union and its client states to also join in this rhetoric and to normalize the idea of peace and humanity in space rather than conflict and enmity. This attempt was, in general, successful at the surface level, though the content of these norms would be contested.

As the United States succeeded in its attempt to push this supranational ethic for space activities into the international system, the "benefit of all mankind" began to alter the dynamics of international norms,[14] which is to say: the globalizing nature of *Sputnik* was not just part of an ongoing techno-military tête-à-tête between the United States and the USSR. *Sputnik* quite literally opened up new spatial realities within an international system established by the UN Charter, which had recently adopted a new territorial settlement at the world scale. *Sputnik*, then, was not just destabilizing in terms of arms parity; it was destabilizing in that it created a possible rift in the new, explicitly anti-imperial system established with the UN Charter.[15] As a result, the newly constituted international community turned its attention to incorporating space into the international geography. This was done by declaring that space and celestial bodies could not be the territory of states or subject to claims of sovereignty.[16]

The international community also took up the idea of linking space to the supranational "benefit of all mankind" in order to heighten the legal obligations that states would have within that new territory. This move distinguished space from other "global commons" such as Antarctica and the high seas, which lack a similar principle governing their exploration and use.[17] This link can be observed in the progression of the term in UN General Assembly (UNGA) resolutions. The first UNGA resolution dedicated to space activities, "Question of the Peaceful Use of Outer Space," was adopted in December 1958.[18] It opens its preamble text with "recognizing the common interest of mankind in outer space."[19] Later in the preamble, the UNGA notes that it desires to "promote energetically the fullest exploration and exploitation of outer space for the benefit of mankind."[20] The following year the UNGA adopted a similar resolution that opened with the same language and expanded on the concept by stating the belief "that the exploration and use of outer space should be for the betterment of mankind and to the benefit of States irrespective of their economic or scientific development."[21] This brings together the ideas that space should be used for "mankind" and that those benefits flow through the nation-state. Nation-states are in turn conceived of as coequals in terms of receiving those benefits. The UNGA did not pass a resolution on space in 1960, and the 1961 resolution adopted the 1959 language verbatim.[22] In 1962, the phrase makes it into the operative text of the resolution and links it directly to the benefit of space communication by stating that the UNGA "believes that communication by satellites offers great benefits to mankind."[23]

In 1963, the UNGA adopted Resolution 1962 (XVII), "Declaration on the Legal Principles Governing the Activities of States in the Exploration and Use of Outer Space." This resolution is significant because it purports to adopt "legal principles," which implies some sort of binding obligation on the state, and it was adopted by consensus of all UNGA members.[24] This cohesion is important because of the nature of the content of this resolution. It begins by repeating the "common interest of mankind" and "betterment of mankind" language in its preamble. After the preamble, the text articulates nine principles to "guide" states in their space activities, the first of which is:

> The exploration and use of outer space shall be carried on for the benefit and in the interests of all mankind.[25]

Not only has the obligatory nature of the principle been heightened in this resolution, but the addition of "all" seems to be an indicator of universality of a right to these benefits. The principles contained in this declaration would become the groundwork for the principles adopted in the 1967 Outer Space Treaty, which pushes the term forward in interesting ways.

The Outer Space Treaty includes the "common interests of all mankind" in the preamble. Then, the first operative phrase of the treaty states:

> The exploration and use of outer space, including the moon and other celestial bodies, shall be carried out for the benefit and in the interests of all countries, irrespective of their degree of economic or scientific development, and shall be the province of all mankind.[26]

This legal obligation changes the principle. Benefits flow to states as representatives of mankind, making the state the responsible intermediary, despite the dedication of space as a "province of mankind," which implies some sort of equality of access. The treaty furthers this notion in Article V when it declares astronauts to be the "envoys of mankind," indicating a duty to represent the interests of all humans and not just those of their respective states.[27] In effect, a state must pursue benefits for all mankind, even those outside its border, but it is also the distributor of the benefits from space to the mankind within its borders. This divergent obligation is indicative of the schism between international human rights and domestic implementation, which is discussed in chapter 3.

As a matter of law, these were new and innovative principles, but they are wholly ambiguous in meaning. The United States' project of attempting to normatively link space exploration to a high-level concept of mankind came to fruition in the Outer Space Treaty, but in practice that connection has been a weak one in terms of actual benefit sharing. While the norm itself drew from the emerging international human rights regime, the prospective bridge between outer space and humankind was never completed. This is in large part because the United States deployed it as a mechanism for pursuing security rather than human rights. The gap between the ideal of providing benefits to all mankind and the actual implementation of the norm is fueled in part by the ambiguity in the norm, which can be read in either liberal, capitalist terms or in socialist, Communist terms. While this was a necessity in attaining state consent to the Outer Space Treaty, which would have been a dead letter without signatures from both the United States and the USSR, it also highlights the deeper rift in the international system on what is owed to "mankind" through the international system. This debate, of course, has deep historical and theoretical roots,[28] and while the UN Charter's dedication to human rights was a watershed moment in world governance, the content of those rights is still deeply contested. Central to this problem is that the 1945 settlement, while bringing the concept of the "human" into the international governance system, pulls forward territorial sovereignty as the core legitimating feature within the system.[29] Though the idea of the "benefit of all mankind" runs throughout the discourse of space governance, by the time it is memorialized in the Outer Space Treaty, those benefits are to be distributed to the states, and the states in turn owe an ambiguous obligation to mankind.[30] Thus, the treaty repeats the Space Age trajectory of *homo americanus*: cohered within the nonterritoriality of space, the supranational ethic "of all mankind" is reflexively reterritorialized through the nation-state's stewardship of its "benefits."

Mankind in the US Space Age

If the state is the mediator of the benefits that are intended to flow to mankind from space operations, then it is proper to look to a state's space program to understand how that state conceived its duty to use space for the benefit of mankind. In the context of the United States, this places the focus on NASA as the civil branch of the American space program, which was established statutorily for the "benefit of mankind."

The Space Race of the 1960s is steeped in the ideological tug-of-war of the Cold War, as both superpowers sought to achieve firsts that symbolically displayed the superiority of their respective governance systems.

This, of course, culminates with the Apollo 11 Moon landing in July 1969, which marks the high point in the Space Race. The United States carefully framed this event in terms that continued to reinforce the "benefit to mankind" norm. Neil Armstrong's first words upon setting foot on the lunar surface invokes his role as an "envoy of mankind," thereby asserting representation of all humans when he breaches new territory.

The American flag that the astronauts planted on the Moon also serves as a potent symbol. This act, on the one hand, is meant to evoke territorial expansion as understood in the post-Westphalian imperial system, in which European claims to new territory were signified by planting the flag of the sovereign as symbolic claim to the soil beneath.[31] On the other hand, the United States intends to dispel that notion with a plaque attached to the ladder of the lunar lander that states, "We came in peace for all mankind."[32] The American flag then represents not a claim of territoriality but a claim of universal ideology.

The framing of Apollo 11 is meant to at once illustrate the triumph of the United States (and liberal ideology) and to universalize this triumph by framing the United States as the representative of all mankind. This representation is imbued with Western, liberal democratic ideals. The United States effectively inscribes the lunar surface with an ideological triumph that is meant to be evoked whenever humankind turns its gaze toward the Moon. While Apollo's disclaiming of territorial sovereignty complies with the anti-imperial mandate of Article II of the Outer Space Treaty, it is not correspondingly noncolonial. The flag is a representation of ideological colonialism. It is not just that the United States "won" the Space Race, but it successfully colonized its ideology onto the lunar surface and into a supranational sphere. This event was carefully scripted to convey a specific international political content, but it is not devoid of domestic political content. The Moon landing's signaling is twofold, then: the plaque disclaims imperial rights, but the flag represents a colonial moment.

Mankind in the Long Space Age

While the civil rights movement is dealt with in detail in chapter 3, it should be noted here that the civil rights movement and the Space Age have significant temporal overlap. *Brown v. Board of Education* was decided in 1954, and the Eisenhower administration committed to launching a satellite for the IGY in 1955. *Sputnik I* was launched two weeks after President Eisenhower signed the executive order integrating Little Rock's Central High School consistent with the *Brown* decision. The dominant narratives of both the civil rights movement and the Space Age grow in tandem and peak, respectively, with the assassination of Martin Luther King Jr. in 1968 and the Apollo 11 Moon landing in 1969. The temporal coincidence is interesting because both attempt to reinforce the idea of compartmentalized, closed historical moments with epistemic political meaning. These epistemic moments serve to simplify history into short-hand that encapsulates an idea with a few uppercase words (Civil Rights, Space Age) but undermines the complexity and open-endedness of both projects. For instance, Hall argues that the dominant narrative of the civil rights movement "distorts and suppresses as much as it reveals."[33] Her argument claims that the historical focus on the civil rights movement of the 1960s obscures the longer, more enduring narrative of civil rights across time in the United States, a phenomenon she calls the "long civil rights movement." The long movement stretches into the past as histori-cally rooted and into the future as an incomplete possibility. Significantly, she argues that this framing is used to depict the civil rights movement as a victory for formal rights, thereby suppressing deeper narratives that focus on economic rights.

Similar distortion and suppression can be seen in the narrative of the Space Age, which is often compartmentalized as the race from *Sputnik I* to the Moon landing. This compartmentalization ignores the "long space age" in favor of the United States' domesticated version of the suprana-tional ethic. Closing the Space Age with the Apollo 11 landing frames the American space program as a triumph for Western liberalism acting on behalf of all mankind. At the same time, it suppresses historical connec-tions to strategic nuclear warfare and presents the US space program as one of enduring primacy indefinitely into the future. Similar to Hall's long civil rights movement, we argue for understanding of a "long space age," one detached from the extension of NASA across physical distance and that instead focuses on the long-term project of bringing benefits of space activities to humankind.

The critique of NASA detailed above and in chapter 3 is salient but in-complete. It would be simplistic to view the US space program in terms of Apollo 11 or to limit it to NASA. Just as the idea of "civil rights" would de-velop across the long civil rights movement, the "benefit of all mankind" would continue to develop in the context of space exploration. The Apollo program concluded with Apollo 17 in 1972. This lander also has a plaque attached to its ladder that states, "May the spirit of peace in which we came be reflected in the lives of all mankind."[34] Rhetorically, this relinquishes some of the claim to global representation asserted by the Apollo 11 phrasing. It also reflects a drop-off in the use of the phrase "benefit of all mankind" at the presidential level. President Johnson uses the phrase in relation to space seven times, Nixon two times, Ford zero times, Carter three times, and Reagan two times; all presidents thereafter do not use it.[35] Interestingly, this shift in administrative rhetoric tracks with the conservative co-option of the civil rights movement observed by Hall.[36] It seems to point to the notion that as victors of the Space Race, the United States no longer needed to lean as hard on the normative role of "mankind" for supranational framing of space exploration. This could either be because Apollo 11 can be read as an "end of history"–type moment in which liberalism is extended to the Moon through US representation of humankind or because, as victor, the United States no longer needed to emphasize the norm to restrain the USSR. Regardless of why, the phrase begins to erode as a mechanism of foreign policy.

Despite the retraction of the phrase as framing language at the highest levels of government, the US space program was in fact being used to bridge the distance between the supranational and domestic deployments of "mankind." Internationally, the United States would use the space program to deliver benefits that could be shared and distributed across the human population. The United States was a primary mover in establishing the International Telecommunications Satellite Organization (INTELSAT), an initiative that began in 1964. The 1973 treaty of this organization uses the "benefit of all mankind" language in its preamble and states as "its prime objective the provision, on a commercial basis, of the space segment required for international public telecommunications services of high quality and reliability to be available on a non-discriminatory basis to all areas of the world."[37] One of INTELSAT's primary goals is extending access to global communications to developing nations. Similarly, NASA would develop the Landsat program (later transferred to the National Oceanic and Atmospheric Administration [NOAA]), which would adhere to the principle of "nondiscriminatory access" by states to the remote sensing imagery that the satellite collected as a way to ensure that developing nations could obtain this benefit.[38] Global weather data and GPS also serve as similar examples of US space activities that are distributed to human populations globally and without discrimination. These programs are certainly part of the "soft power" used by the United States to reinforce its conception of liberal values at the global scale, but they also close the distance between the benefits of space and humankind.

NASA also changed within the context of the "long space age," making a post-Apollo turn that provides neoliberal responses to the kind of critiques levied against it by black freedom activists as detailed in chapter 3. Kim McQuaid recounts NASA's initial false start at addressing discrimination in its ranks with the hiring and firing and rehiring of Ruth Bates Harris.[39] This fiasco, though, led to the beginning of a genuine transition at NASA, as Harriet Jenkins began a long stint heading NASA's Equal Opportunity Office. Jenkins emphasized to the decentralized NASA centers that diversity was part of NASA's "mission." By placing representation politics within the bounds of the "mission," Jenkins helped to shift NASA's vision from looking only outward at distant destinations to engagement with populations within its locality.

Jenkins's work is consistent with NASA's neoliberal, post-Apollo turn. As the 1980s approached, NASA moved from the Apollo program to the space shuttle program. The change from old to new was matched with signaling by NASA that its mission had changed. The Apollo and pre-Apollo era was marked by programs named after figures from Greek mythology. Mercury, Gemini, and Apollo are more than just fanciful names for these programs. They are symbols of classical strength and virility, and as the names of the flagship space programs of the 1960s, they signaled the power of the United States in its race with the Soviet Union. While these names may have been chosen to connote strength to the Western ear, they also reinforced the white, masculine face of the space program and its astronaut corps. The space shuttles, however, were named after famous ships of exploration: *Enterprise, Columbia, Challenger, Discovery, Atlantis,* and *Endeavour.* At least four of these names also have connotations of American values, though it should be noted that while these names mark a change from Greek mythology, some of them do reference imperial and colonial exploration (in particular, *Columbia*).

The change in space transportation would also be marked by the integration of the astronaut corps. The class of 1978 astronaut corps included the first female, black, and Asian American astronauts.[40] Nineteen eighty-three saw both the first female astronaut and the first African

American astronaut fly into space on shuttle missions.[41] The shuttle would also open the possibility of carrying international crew members, the first being West German Ulf Merbold on STS-9, also in 1983. The diversifying of the astronaut corps meant that the "envoys of mankind" that NASA sent to space could now evoke *homo americanus* in a visually composite, multicultural frame.

Starting in the 1990s, NASA's institutional pivot toward locality, diversity, and engagement gained a new focus: science, technology, engineering, and math (STEM) education. In 1992, NASA adopted its first agency-wide education policy, which stated that "it is NASA's policy to use its inspiring mission, its unique facilities, and its specialized workforce to conduct and facilitate science, mathematics, engineering, and technology education programs and activities."[42] Redomesticating the supranational ethic once again, these educational programs work to remediate the historical disjuncture between space exploration's "benefit of all mankind" as envisaged in the Apollo era and communities—particularly communities of color—proximate to NASA centers. Space exploration, in this institutional role, is not so much an envoy *of* as *to* humankind—localizing the nonterritorial. For example, the NASA Explorer Schools Program "immerses selected high-minority and high-poverty urban and rural middle schools in NASA mission content by providing them access to NASA resources, people, and products."[43] Similarly, the Minority University Research and Education Program (MUREP) "addresses educational issues of underserved and underrepresented students at both K–12 and higher education levels, through activities undertaken by minority universities."[44] Now a half-decade since Apollo, the benefits of the Space Age continue to provide a context for invoking a national community underwritten by the supranational aspirations of "all mankind" in the US imagination.

Conclusion

This chapter has sought to briefly connect the United States' deployment of human rights rhetoric to normalize and securitize space activities at the national level in the early days of space exploration. This is, of course, only half of the story. As alluded to, this use of human rights rhetoric has direct links to the ongoing struggle for black freedom in the United States. The civil rights narrative is picked up in chapter 3 in order to give a fuller account of NASA's relationship to humankind.

Notes

1. See, for example, Shirley Thomas, *Men of Space*, vol. 7, *Profiles of the Leaders in Space Research, Development, and Exploration* (New York: Chilton Books 1965) 62, 244; Nathan C. Goldman, *American Space Law: International and Domestic* (Iowa City: University of Iowa Press, 1988), 4 ("The event may have been as traumatic as any in American history."); and John M. Logsdon, ed., *Legislative Origins of the National Aeronautics and Space Act of 1958 : Proceedings of an Oral History Workshop : Conducted April 3, 1992* (Washington, DC: Government Printing Office, 1998), vii.

2. White House, "Statement by James Hagerty," July 29, 1955, https://www.eisenhower. archives.gov/research/online_documents/igy/1955_7_29_Press_Release.pdf; National Science Foundation and National Academy of Science, "Plans for Construction of an Earth Satellite Vehicle Announced," July 29, 1955, https://www.eisenhower.archives.gov/ research/online_documents/igy/1955_7_29_NSF_Release.pdf; Thomas, *Profiles of the Leaders in Space Research*, 144.

3. In a meeting after the *Sputnik* launch, President Eisenhower was informed that the US Army felt that it had been in a position to orbit a satellite for months, but that the desire to maintain the civil nature of the United States' IGY participation had prohibited the army from launching. Memo on Sputnik, Memorandum of Conference with President Eisenhower, October 8, 1957, https://catalog.archives.gov/id/186623.

4. Official White House Transcript of President Eisenhower's Press and Radio Conference #123, October 9, 1957, 4, https://www.eisenhower.archives.gov/research/online_ documents/sputnik/10_9_57.pdf.

5. Dwight Eisenhower, "Atoms for Peace," December 8, 1953, https://www.eisenhower. archives.gov/all_about_ike/speeches/atoms_for_peace.pdf.

6. Searching "benefit of all mankind" at the American Presidency Project, http:// www.presidency.ucsb.edu/ws/index.php, renders no results before Eisenhower's "Atoms for Peace" speech. The phrase "benefit of mankind" brings up a number of usages before this, significantly including three usages by President Truman in relation to atomic energy.

7. Dwight Eisenhower, Message to the First Conference of the International Atomic Energy Agency, October 1, 1957, http://www.presidency.ucsb.edu/ws/index. php?pid=10917.

8. Dwight Eisenhower, Statement by the President on Releasing the Science Advisory Committee's "Introduction to Outer Space," March 26, 1958, http://www.presidency. ucsb.edu/ws/index.php?pid=11331.

9. Dwight Eisenhower, Annual Message to the Congress on the State of the Union, January 7, 1960, https://www.eisenhower.archives.gov/all_about_ike/speeches/1960_ state_of_the_union.pdf.

10. Dwight Eisenhower, Proclamation 321—United Nations Human Rights Day, 1957, December 7, 1957, http://www.presidency.ucsb.edu/ws/index.php?pid=107184.

11. Ibid.

12. Ibid., Sec. 102.

13. This took place in November 1958. Logsdon, *Legislative Origins*, 35. Significantly, Johnson, who carefully crafted and shepherded the Space Act through Congress, was also focused at the time on civil rights legislation. Ibid., 58.

14. On international norm development, see, generally, Martha Finnemore and Kathryn Sikkink, "International Norm Dynamics and Political Change," *International Organization* 52, no. 4 (1998): 887–917.

15. The UN Charter adopts the "self-determination of peoples" as one of its purposes. UN Charter, Art. 1.2. The anti-imperial approach can be observed in the Outer Space Treaty's treatment of sovereignty and appropriation in Article II. See P. J. Blount and Christian J. Robison, "One Small Step: The Impact of the U.S. Commercial Space Launch Competitiveness Act of 2015 on the Exploration of Resources in Outer Space," *North Carolina Journal of Law & Technology* 18 (2016): 164.

16. UN Res. 1962 (XVIII), Declaration of Legal Principles Governing the Activities of States in the Exploration and Use of Outer Space (December 13, 1963) and Treaty on Principles Governing the Activities of States in the Exploration and Use of Outer Space, Including the Moon and Other Celestial Bodies, 610 UNTS 205 (October 10, 1967), Art. II (hereafter cited as Outer Space Treaty).

17. See Antarctic Treaty, 402 UNTS 71 (December 1, 1959), and United Nations Convention on the Law of the Sea, 1833 UNTS 3 (December 10, 1982).

18. An earlier mention of space activities was made in a 1957 resolution on disarmament, which requested that study be done on a system to "ensure that the sending of objects through space shall be exclusively for peaceful and scientific purposes." UN Res. 1148 (XII), Regulation, limitation, and balanced reduction of all armed forces and all armaments; conclusion of an international convention (treaty) on the reduction of armaments and the prohibition of atomic, hydrogen and other weapons of mass destruction (November 14, 1957).

19. UN Res. 1348 (XIII), Question of the Peaceful Use of Outer Space (December 13, 1958).

20. Ibid.

21. UN Res. 1472 (XIV), International Co-operation in the Peaceful Uses of Outer Space (December 12, 1959).

22. UN Res. 1721 (XVI), International Co-operation in the Peaceful Uses of Outer Space (December 20, 1961).

23. UN Res. 1802 (XVII), International Co-operation in the Peaceful Uses of Outer Space (December 14, 1962). Interestingly, this phrase is not used in the portion of the resolution on meteorology from space.

24. As a general rule UNGA resolutions do not create binding international law, but the nature of the assertion of "legality," coupled with "consensus," has led to the idea of "instant customary international law." For our purposes, regardless of how binding these principles are, they do illustrate the progress of the idea of the "benefit of all mankind" down the path to hard legality. On instant customary international law, see Bin Cheng's classic article, "United Nations Resolutions on Outer Space: 'Instant' International Customary Law?," *Indian Journal of International Law* 5 (1965): 23.

25. UN Res. 1962 (XVIII), Declaration on the Legal Principles Governing the Activities of States in the Exploration and Use of Outer Space (December 13, 1963).

26. Outer Space Treaty, Art. I.

27. Ibid., Art. V.

28. See generally Micheline Ishay, *The Human Rights Reader: Major Political Essays, Speeches and Documents from Ancient Times to the Present*, 2nd ed. (New York: Routledge, 2007).

29. "Human Rights" are addressed in Article 1 of the UN Charter, but are subordinate to "international peace and security." Article 1.1 characterizes the obligation of international peace and security with verbs such as "to maintain" and "to take effective collective measures." Article 1.3, on the other hand, uses the verbs "promote" and "encourage" to describe "human rights" obligations. UN Charter, Art. 1.

30. This reading is confirmed by Articles VI–IX of the Outer Space Treaty, which, read together, make states the primary stakeholders in outer space. Outer Space Treaty, Arts. VI–IX.

31. In the US Senate hearings on the Outer Space Treaty, Senator Frank Church asked if Article II meant that space and celestial bodies "cannot be claimed for Ferdinand and Isabella." Ambassador Arthur Goldberg responded that this interpretation "is correct." Thomas Gangale, "The Legality of Mining Celestial Bodies," *Journal of Space Law* 40 (2015–2016): 199.

32. NASA, "Apollo 11 Plaque Left on the Moon," https://www.nasa.gov/centers/marshall/moonmars/apollo40/apollo11_plaque.html.

33. Jacquelyn Dowd Hall, "The Long Civil Rights Movement and the Political Uses of the Past," *Journal of American History* 91 (March 2005): 1233.

34. NASA, "Apollo 17 Plaque," https://solarsystem.nasa.gov/galleries/apollo-17-plaque.

35. Searching the American Presidency Project, http://www.presidency.ucsb.edu/, for "benefit of all mankind." Interestingly, this phrase is used by presidents to describe myriad other international issues.

36. Hall, "Long Civil Rights Movement," 1248–1250.

37. Agreement relating to the International Telecommunications Satellite Organization "INTELSAT," Art. III (August 20, 1971).

38. 15 CFR 960.12 (2017). See also Joanne Irene Gabrynowicz, "One Half Century and Counting: The Evolution of US National Space Law and Three Long-Term Emerging Issues," *Harvard Law and Policy Review* 4, no. 2 (2010): 405, 416; and Joanne Irene Gabrynowicz, *The Land Remote Sensing Laws and Policies of National Governments: A Global Survey* (National Center for Remote Sensing, Air, and Space Law, University of Mississippi School of Law, 2007).

39. Kim McQuaid, "Race, Gender, and Space Exploration: A Chapter in the Social History of the Space Age," *Journal of American Studies* 41, no. 2 (2007): 405–434.

40. NASA, "1978 Astronaut Class," https://www.nasa.gov/image-feature/1978-astronaut-class.

41. As a side note, a woman, Peggy Whitson, currently holds the American record for the most time spent in space. Claire Zillman, "Astronaut Peggy Whitson Has Now Spent

More Time in Space Than Any Other American," *Fortune*, April 24, 2017, http://fortune.com/2017/04/24/nasa-peggy-whitson-astronaut-most-time-space/.

42. Helen R. Quinn, Heidi A. Schweingruber, and Michael A. Feder, eds., *NASA's Elementary and Secondary Education Program: Review and Critique* (National Academies Press 2008), 23, http://www.nasa.gov/pdf/550499main_Elem-Sec-EdProg-Review-Critique.pdf.

43. Ibid., 13.

44. Ibid., 35.

Bringing the Moon to Mankind

The Civil Rights Narrative and the Space Age

DAVID MIGUEL MOLINA AND P. J. BLOUNT

In June 1969, the Poor People's Campaign planned a march from US senator James Eastland's plantation in Sunflower County, Mississippi, to Florida's Cape Kennedy. Coming a year after the attempt by the Southern Christian Leadership Conference (SCLC) to mobilize a radical, multiracial coalition of the nation's poor through a monthlong encampment in the National Mall in Washington, DC, the Cape Kennedy march—as part of the campaign's "hunger phase"—dramatized a then-familiar critique of federal policy priorities in the face of enduring poverty conditions in urban and rural communities nationwide. Aiming "to protest the spending of money on space joyrides while children are dying from hunger and starvation every day," the journey culminated just days before Apollo 11 was to launch for the first human Moon landing.[1] Upon arriving at the launch facility on July 15, the march's organizer, Rev. Ralph Abernathy, was greeted by National Aeronautics and Space Administration (NASA) administrator Thomas O. Paine, who assured the SCLC leader that "if we could solve the problems of poverty by not pushing the button to launch men to the moon tomorrow, then we would not push that button." As a concession, however, Paine was able to extend Abernathy and his fellow marchers—a group of over 150, comprising twenty-five low-income families from five southern states—forty VIP passes to watch the Apollo 11 launch.[2] Abernathy accepted, welcoming the opportunity to address the crowd gathered at the viewing site, lead freedom songs with Poor People's Campaign activists, and sign autographs.[3] And so, as *Jet* magazine's Chester Higgins remarked, some of the protesters were able to enjoy "choice seats from which to watch the millions of dollars that might have been spent on their welfare disappear in a thunderous rocket blast as Apollo 11 soared off into the wild blue yonder."[4] Having come to contest the Moon shot as evidence of their marginality from the interests of state power, the nation's poor found themselves incorporated into the event as spectacular participants.

This historical vignette is more than just a momentary overlap in the history of the civil rights movement and the US space program. Instead, it is one of numerous cultural, social, and political episodes that illustrate the critical disjuncture between NASA, the US space enterprise, and the struggle for black freedom. The Cape Kennedy march is of particular importance as it is enacted in a conjunctural moment for a number of discursive, conceptual, and administrative tensions that illuminate NASA's situatedness within the long civil rights movement. As such, the Poor People's Campaign protest has played a visible—though limited—role in both media coverage of the Apollo 11 launch and in subsequent scholarship on both the Space Age and post–civil rights challenges to develop more substantive critiques of enduring economic, social, and political inequality in the aftermath of legislative victories in 1964 and 1965.[5]

In returning to the Cape Kennedy protest, then, this chapter is not posing an investigation into whether criticisms of the Apollo program were voiced at all by contemporaneous social movements or the black press. Indeed, as Roger Launius suggests, concerns about manned space flight as an "embarrassing national self-indulgence" were widely common in the public at large across the arc of the Apollo program—an unease in the national mood set just beneath the otherwise triumphalist reception of the Moon landing itself. And, as Lynn Spigel explores in *Welcome to the Dreamhouse*, this disjuncture was only amplified in coverage by black media outlets. In many articles, the title alone tells the tale: for example, "Moon Probe Laudable—But Blacks Need Help" and "Blacks and Apollo: Most Couldn't Have Cared Less."[6]

A critique of the Moon landing would have contemporary articulations in black cultural production as well. Released a year after the Apollo 11 launch, Gil Scott-Heron's debut album, *Small Talk at 125th and Lenox*, featured the track "Whitey on the Moon." Recorded live, Scott-Heron performs a poem over accompanying percussion. Initiating with the disclosure that "a rat done bit my sister Nell / with Whitey on the moon," the composition oscillates between an entangled portrait of economic dispossession—Nell's medical needs; unsafe living conditions; increases in rent, food prices, and taxes—with the recurrent, matter-of-fact observation that "Whitey's on the moon."[7] As the track concludes, this juxtaposition is bridged by a gesture of powerlessness: without money left to address basic needs, Scott-Heron resolves to simply send "these doctor bills / Airmail special / to Whitey on the moon." Here, as in many of the responses in black popular culture at the time, the self-evident triumphalism of Apollo 11 is rendered an absurdity. With each repetition of "Whitey on the moon," the phrase's descriptive content increasingly hollows out an ultimately incoherent act in the face of everyday struggle.

In short, the Poor People's Campaign was working within a discursive terrain in which critiques of the Moon shot—especially in black communities throughout the country—were the rule, not the exception.[8] Complementing our chapter in this volume about the rhetorical fissures proved by articulations in US space policy of space exploration's "benefit for all mankind," this chapter is concerned with how the Cape Kennedy protest and other critiques reveal a range of black political imaginaries working to make legible the problem exposed by the Apollo mission's success. In chapter 2, we traced the evolution of a supranational ethic, "for the benefit of all mankind," and its journey to the surface of the Moon. Here we return to Cape Kennedy to explore how the Moon shot—as a spectacular event—allowed for various strategies in pivoting this notion "of all mankind" from aspirational ethos to insurgent political demand.[9]

The Wagon, the Rocket, and the Slave Ship: The Lunar Topos in Black Political Imaginaries

Critiques of the Moon landing from within black freedom initiatives serve as a useful tool for exposing gaps within human rights discourse proliferated by the US space program. In framing the Apollo missions as spectacular disjunctures between state motives and the needs of black citizens, black freedom critiques mark different inflections in the "long" resonance between Space Age and post–civil rights discourses: from Great Society claims to liberal democratic fulfillment under the aegis of technocratic and social progress, to the coeval Cold War struggle over the fraught availability of "human rights" as a supranational ethos. With the Moon shot, the varieties of this struggle play out as a journey to the interstices of state sovereignty and nationhood—as state policy and freedom struggle mobilize the Moon landing to conceptualize a human rights regime that simultaneously exceeds and extends national interest.

The official US policy forwarding space exploration and use "for the benefit of all mankind," examined in chapter 2, provides the rhetorical grounds for a range of these dissenting rearticulations of the Moon landing. In the discussion that follows, we look at two: the Poor People's Campaign protest discussed in the introduction, and a series of illustrations produced by Emory Douglas of the Black Panther Party. Though each mobilizes Apollo to critique the US state, their mobilizing political imaginaries are rather distinct—and mirror long-standing tensions in black radical, liberal, and civil rights formations in the United States.[10] In pushing back on the Moon shot spectacularly professing to "come in peace for the benefit of all mankind," this tension is set in high relief: whereas the Poor People's Campaign demanded a corrective to the unevenness of the state's peace and benefits, the Black Panther Party will call for a wholesale dismantling of the colonizing premise animating the notion of "mankind" itself.[11]

Yet these otherwise divergent critiques share a common rhetorical framing: what we call here the lunar topos. A key analytic of rhetorical criticism, topoi are commonplace structures for argument—historically situated modes of claims-building that bind together speaker and audience through assumptive familiarity. Regarding the availability of a "lunar" topos in the "long space age," we suggest that the sublime gap between extraplanetary voyage and terrestrial concern produces a common framework for deploying the "lesson of Apollo" in discourse,[12] which is to say, progressive arguments of the form "If we can go to the Moon, then we can do X" abound in public culture—as do their critical obverse, "If we can go to the Moon, then why haven't we done X?" The back-and-forth between these claims—in which the Moon landing is simultaneously symbolic of a possible future and foreclosed present—is the lunar topos.

By bringing the Poor People's Campaign to Launch Complex 39, Ralph Abernathy deployed the lunar topos via protest. Insistent that he and fellow protestors get "as close as possible to the space capsule when it is launched," the SCLC leader and fellow organizers worked to dramatize the Space Age as one of stasis, not progress, for US blacks.[13] Central to this performance was the use of anachronism—enacting the lunar's critical edge by juxtaposing the Apollo rocket with a reprise of the Poor People Campaign's Mule Train caravan.[14] The Mule Train had been developed for the campaign's initial phase in the summer of 1968—as one of eight caravans from across the country set to converge on a shantytown constructed in the National Mall for the purposes of hosting a multiracial coalition of the nation's poor. The project—an experiment in "militant nonviolence" seeking to forge a new civil rights consensus around economic dispossession with New Left, Black Power, Chicano/a, American Indian, and poor white projects—would be Martin Luther King Jr.'s last major endeavor, still in the final stages of preparation at the time of his assassination.[15]

Under the new leadership of Ralph Abernathy, the SCLC and the Poor People's Campaign persisted its inaugural project after King's death—and on May 13, 1968, the Mule Train caravan left the Delta town of Marks, Mississippi, en route to Washington, DC. It is important to note, however, that even in this initial iteration the Mule Train was carefully constructed as a performance of Space Age anachronism. Quite simply, no one was using mule-drawn wagons in the Mississippi Delta in 1968—it took a substantial organizing effort to even locate enough mules to pull the caravan.[16] As such, the protest vehicle cum photographic spectacle played on the general public's notion of the underdevelopment of rural southerners and impoverished blacks to dramatize real inequalities in unemployment, social services, and education. This visual bait-and-switch was reinforced by messages painted on the canvasses covering the mule-drawn carriages: "Don't laugh, folks: Jesus was a poor man," "Everybody's Got a Right to: Work, Eat, Live," and, notably, "Which is Better? Send a Man to the Moon or Feed Him on Earth?"[17]

In revisiting the Mule Train strategy a year later, the Poor People's Campaign amplified the anachronistic potential in this last slogan to its presumable limit: the juxtaposition of "a gigantic moon rocket and an old mule-drawn wagon [writing] a paradox of humanity," as a reporter for the *Boston Globe* put it.[18] And, once again, this performative lunar topos mobilized a visual suspension to circulate a linked political demand: addressing widespread hunger and malnutrition in cities and rural areas as effects of poverty. If we can "send a man to the Moon," the Mule Train demanded, then why can't we "feed him on Earth?" These concerns were broadcast through every aspect of the protest: from the departure at Senator Eastland's Delta plantation—highlighting the discrepancy between the thousands of dollars of federal farm subsidies issued for the senator to "not grow food or fiber" on his land and the average "$8.50 a month in welfare allotments" distributed to eligible Mississippi children—to the arrival at Cape Kennedy—presenting what Abernathy called "the 51st state of hunger," marching alongside the Mule Train with signs reading "Billions for Space, Pennies for Hunger," "Moonshots Breed Malnutrition," and "Rockets or Rickets?"[19]

In meeting with NASA's Thomas Paine, however, the Poor People's Campaign confronted a familiar mechanism of state accommodation: a US government simultaneously triumphant and overwhelmed. In the face of the complexities of widespread hunger under conditions of poverty, Paine characterized the otherwise massive technological and managerial feat of landing a human on the Moon as a mere "push of a button." Nevertheless, in a gesture reminiscent of decades of liberal deflection in the face of civil rights demands, NASA's chief resolved the campaign's dramatization of the lunar topos by inverting it: asking the protesters to "hitch your wagon to our rocket and tell the people the NASA program is an example of what this country can do."[20] Though they came seeking an overhaul of federal antipoverty programs, Abernathy and the others found themselves in a familiar civil rights limbo of deferred demands—granted equality of audience to state power, but hardly a serious consideration of their challenge to reframe the state's audience to their powerlessness. As mentioned earlier, in accepting Paine's invitation of VIP tickets to view the Apollo 11 launch, Abernathy attempted to leverage the visibility of the event to publicize the Poor People's Campaign. However, the spectacular nature of the Moon landing meant that any intended associations would be hopelessly upstaged. In comments to press afterward, the SCLC leader even suggested that as he "saw the rocket take off . . . the pride that I felt for it made me forget momentarily the two worlds of America."[21] Through formal incorporation into the event, Abernathy and his fellow Poor People's Campaign activists had become participants in their very object of protest, special guests on the sidelines of "what this country can do."

For Emory Douglas of the Black Panther Party, however, the rocket itself was always the concern. As portrayed in a pair of illustrations in the biweekly *Black Panther Community News Service*, Douglas likewise presented a critique that linked the Apollo missions to histories of slavery, sharecropping, and rural unemployment. However, for Douglas, the simultaneity of the Moon shot and conditions of black dispossession occasioned a rejection of these overlapping iterations of state power, not a call for that power's corrective capacity. In short, if the Poor People's Campaign protest deployed the lunar topos to bring the plantation to the Moon landing—if NASA can land on the Moon, then why can't the US state eradicate poverty?—Douglas's illustrations brought the Moon landing to the plantation—if NASA can land on the Moon, then what new modes of power are available to the state?

In two issues of the Black Panther Party's communications platform— July 26 and August 23, 1969—Douglas used the recent Moon landing as the setting for full-page illustrations exploring the "lesson of Apollo" for the paper's circulation audience of 140,000.[22] Images like this were a common feature in the *Community News Service*. By then, illustrations by Douglas, the party's minister of culture, were a crucial element of the paper's layout, bookending accounts of party activity, speeches from leaders, updates on incarcerated members, reports from party branches throughout the country, and news from anticolonial struggles and radical allies at home and abroad. As the scholar of black visual culture Leigh Raiford notes, Douglas's broadsheet images "proved critical to the emergent visibility of the [Black Panther Party], translating the party's ideology and rhetoric into single frame cartoons" that functioned as readymade posters.[23] In giving the Moon shot the full-page treatment, then, Douglas sought to translate the lunar topos within the visual logics of the Black Panther program.

Like the Mule Train caravan protest, the core visual strategy in Douglas's illustrations hinges on anachronism. However, in this instance the tension is interior to Apollo itself—not set between the Space Age rocket and an anachronistic foil. In both of Douglas's images, the rocket-as-vessel is narrativized as yet another extension of the colonial relation's long durée.[24] In the first illustration (figure 21, page 108), published as the back page of the *Community News Service* ten days after the Moon landing, Apollo functions as a Space Age slave ship. Rather than a lunar module, Apollo appears here as an upright rocket, its nose in the shape of a pig's head—the Panther sobriquet for an oppressor. At the rocket's base, an open door reveals a chain gang making the small step to the Moon's surface. As one of the human figures remarks, "I knew we should have stop [sic] this shit before it got off the ground." A group of three pig figures oversees the procession. One, handing over a shovel and pickax, notes of the situation, "Hey handle those slaves with care we're gonna to need them for Mars, Pluto, and all the other planets."[25]

A slogan, "Whatever Is Good for the Oppressor Has Got to Be Bad for Us," captions the July 26 image. This, quite succinctly, is Douglas's take on the lunar topos: a reading of "If we can go to the Moon, we can do X" that sees the first part of this conditional not as triumph, but as horror. Here, the notion that whatever benefit comes from Apollo would be distributed "for all mankind" is met with a harsh reminder that the notion of "mankind" itself has a fraught history—as do many initiatives conducted in its name. In connecting manned space flight with the transatlantic slave trade, Douglas produces a visual suspension that rejects outright the possibility that impoverishment and space travel are the uneven effects of misplaced priorities. Rather, Douglas's lunar chain gain suggests that dispossession and progress are coextensive within the basic priorities of the state.

This line of thought is revisited a month later by Douglas in the August 23 issue. Apollo is again depicted as an upright rocket, but in this image only one body is in the frame: a pig-figure wearing a spacesuit with a fishbowl-style helmet (figure 22, page 108). While hammering a sign into the lunar surface that reads "Whites Only," the figure exclaims, "At Last!" Here, the rocket functions not as the slave ship but as an exploratory vessel of colonial conquest. As the only astronomical body to have been traversed by whites alone, here the Moon becomes a dystopic terrain of total racial exclusion. Echoing the national commitments underwriting the supranational ethic "of all mankind," the Apollo 11 plaque and the American flag merge in Douglas's illustration, replaced with the iconic placard of the Jim Crow South.[26] No Black Panther slogan captions this image, but the iconography presents a clear refrain of July's Apollo image: whatever technological advancements it may bring, the Space Age is just another version of colonial power.

Unlike the Poor People's Campaign's deployment of the lunar topos at Cape Kennedy—in which a realignment of state funding might address the gap between "rockets" and "rickets" dramatized by the Mule Train's anachronistic performance of poverty—there is no corrective offered to the mode of governance presented in Douglas's illustrations, and no organizational engagement with NASA as such. By presenting *Mare Tranquillitatis* as a dystopic terrain for black political imaginaries, Douglas's visual response to the Moon shot provides some insight into the fact that future articulations of Afrofuturism in black cultural production were pitched at a decidedly galactic register. For space to be "the place," as it were, it had to be anchored in a political topography outside of the reach of terrestrial incorporation.[27] On the other hand, the Black Panther Party's concrete (and fraught) search for political alternatives addressing similar modes of dispossession protested by the Poor People's Campaign filled much of the rest of the *Community News Service* issues in which the Douglas illustrations appeared. Rather than meeting with NASA administrators or lobbying the Department of Agriculture for expansion of the food stamp program—a legislative victory afforded to the Poor People's Campaign's initial phase in 1968—Black Panther chapters throughout the country were managing free breakfast programs and health clinics.[28]

This is not to claim that Black Panther community survival programs were wholly successful as solutions to widespread malnutrition and lack of access to health care, or that the significant increase in food stamp allocation won by the Poor People's Campaign in 1968 and repoliticized at Cape Kennedy in 1969 was somehow totally inadequate. Rather, these different attempts to ground the "lesson of Apollo" illustrate the range of political imaginaries available to post-1965 black freedom projects—different modes of leveraging the Moon shot to describe the political present and to advance a political future. At the end of the day, what differs here are the projects' respective treatment of the nation-state's role in this dynamic, particularly in how resolutions of the lunar topos might play out for each protest. For the SCLC, the Poor People's Campaign is a call for the US state to fulfill its democratic promise; were the country to adopt different priorities—to redirect the activity of "what this country can do"—the issue of hunger would be better dealt with. For the Black Panther Party, the Emory Douglas illustrations present the state as a failed mechanism outright and the Moon landing as but a new technology of oppression. The challenge to the nation-state here is not pitched at the fulfillment of duties but at rethinking the notion of the state wholesale. For the Panthers, Apollo's lesson of "what this country can do" amplifies the need to build local capacities of self-determination and transnational, radical alliances of shared dispossession. As an article in the October 11, 1969, issue of the *Community News Service* remarks on the collaboration of the Black Panthers and the Young Lords Organization (a radical Puerto Rican community organization) in New York City on the distribution of clothing to children and families suffering from welfare cuts: when the people were served, "the state was nowhere to be seen."[29]

Conclusion: "Mankind" at a Distance

In both the Poor People's Campaign and Black Panther Party's critiques of the space program, the salience of the lunar topos hinges on a conceptualization of distance. The Mule Train traversed physical distance to hyperbolize a figurative, temporal one; the Douglas illustrations portray space exploration as ironically interior to black communities—and productive of those communities' Space Age powerlessness. The use of distance as central logic in these critiques underscores the fraught overlap between "mankind" as understood across articulations of black freedom and "mankind" as understood from within discourses on space exploration. Despite the fact that both the Mule Train and Douglas's illustrations play on the supranational conception of human rights broadcast by the US space program, their different lessons of Apollo evidence a sharp rhetorical divergence in these very grounds.

In bringing the Mule Train to Cape Kennedy, the Poor People's Campaign offered a performance of distance that contested the legibility of NASA's global notion of humankind. These protestors invited NASA to, quite literally, look out its window and to see the dispossessed of "mankind" as a constituent demand on the state's priorities, not merely a spectator to them.[30] The Poor People's Campaign's critique is centrally concerned with the distribution of state "benefits." The complaint is not so much that NASA is de facto racist. Instead, it is that the state needs to bridge the distance between the wagon and the rocket—rethinking the composition and legibility of the "mankind," both locally and globally, in whose interests it is acting. NASA's patronizing response to Abernathy's lunar topos—tickets to see the spectacle—evidenced its conception of the mission and ignored populations that would likely still have in their recent collective memory the spatial displacement caused by the establishment of these government facilities.

In contrast, the critique levied by Douglas leans on a radical rejection of the nation-state as a structure through which benefits are distributed. The problematic composition and legibility of "mankind" underwriting the Moon shot are not the result of an overlooked emphasis, but a stark reification of colonial power. Here, Douglas provides a visual echo of Frantz Fanon's claim in *The Wretched of the Earth* that "in this age of Sputnik we might think it ridiculous to die of hunger, but for the colonized masses the explanation is more down to earth."[31] For Fanon, the management of population is a colonial practice, and space exploration will, as depicted by Douglas, simply be a dissembling extension of these colonial values. The very assumption that there is a choice between space exploration and feeding the hungry reproduces the rationale of power that effects dispossession. For Fanon, Douglas, and other anticolonial Space Age critics, the lunar topos's problematic distance is not a gap to overcome but a proximity to refuse.

As discussed in chapter 2, during the Cold War the "benefit of all mankind" and the "province of all mankind" were operationalized in US policy as ambiguous legal and political terms to frame space exploration as an enterprise entangled with human rights. This ambiguity was quite purposeful. The great powers of the bipolar world were free to imbue "benefits" and "common interests" with either liberal democratic or radical socialist conceptions of rights. The phrase allowed for an air of consensus without commitment to an ideologically untenable norm and is rooted in an agreement that space exploration will not be an imperial venture at the level of sovereign territorial expansion. However, despite this anti-imperial stance, the phrase is not anticolonial; the race to the Moon was very much about colonizing the lunar surface with the ideological values of the victor. In the words of Edward C. Welsh, who served as executive secretary of the National Aeronautics and Space Council, "the outcome of this race may well determine whether or not freedom becomes a second-rate commodity."[32]

While states intended the ambiguity to functionally reside in the first half of the phrase "for the benefit of all mankind"—the interpretation of "benefit" and the allocation of benefits within Cold War international relations—they cast a shadow over the second half, "mankind." As seen in both the Poor People's Campaign and the Black Panther critiques, the latent indeterminacy of the constitution of "man" provides the engine of contestation for the lunar topos. Regardless of the "benefit," whether the Moon landing is a reservoir of progressive optimism or a glimpse at new colonial horror hinges on how the audience is constituted, and how it is situated vis-à-vis the "we" of "If we went the Moon, then . . ." These critiques capitalize on the malleability found at the very core of both the US space program and international space governance: a felt distance between the concept of "mankind" and the actual individuals subject to the allocation of state benefits.[33]

In both instances, NASA was ill equipped to counter these critiques. And yet, even if the agency had wanted to, it would have been impossible to completely overcome the systemic and structural conditions overdetermining the white space of the flagship civil space program.[34] As a supranational ethos, the core ambiguity in "mankind" was rooted in the Cold War rivalry between the United States and the USSR—and NASA was a critical, motivated component of the US counter to global Communism. Though NASA posed its statutory mandate to act for the "benefit of all mankind" in global terms, the mission forwarded a notion of humankind crafted in the image of *homo americanus*. The view from Apollo is one that looks from the Moon back at Earth and asserts "we," not NASA or the US state, have bridged this distance—ever colonizing the former by the latter. As Apollo left Launch Complex 39, NASA's trek heavenward obscured the humankind that surrounded it.

Notes

1. Hosea Williams, memorandum to SCLC Board of Directors, June 27, 1969, Box 285, Folder 2977, Chicago Urban League records, Special Collections and University Archives, University of Illinois at Chicago.

2. "NASA Chief Tells Abernathy His Side," *Atlanta Constitution*, July 16, 1969.

3. "Lift for Poor Urged at Lift-Off," *The Sun*, July 17, 1969.

4. Chester Higgins, "SCLC President Says Greatest Victory Is for Nonviolence," *Jet*, August 14, 1969.

5. For discussions at the time about civil rights moving into a "second phase" after liberal democratic victories in civil rights legislation, see Bayard Rustin's argument in "From Protest to Politics: The Future of the Civil Rights Movement," *Commentary* 39, no. 1 (1965): 25–31, and "The Lessons of the Long Hot Summer," *Commentary* 44, no. 4 (1967): 39–45. For accounts of the Poor People's Apollo march in scholarship, see Roger D. Launius, "Managing the Unmanageable: Apollo, Space Age Management and American Social Problems," *Space Policy* 24, no. 3 (August 2008): 158–165; Brenda Gayle Plummer, *In Search of Power: African Americans in the Era of Decolonization, 1956–1974* (Cambridge: Cambridge University Press, 2013); Lynn Spigel, *Welcome to the Dreamhouse: Popular Media and Postwar Suburbs* (Durham, NC: Duke University Press, 2001); Matthew D. Tribbe, *No Requiem for the Space Age: The Apollo Moon Landings and American Culture* (Oxford: Oxford University Press, 2014).

6. Roger D. Launius, "Public Opinion Polls and Perceptions of US Human Spaceflight," *Space Policy* 19, no. 3 (August 2003): 166. Spigel, *Welcome to the Dreamhouse*, 164; Simeon Booker, "Moon Probe Laudable—But Blacks Need Help," *Jet*, August 1, 1969, 10; Thomas A. Johnson, "Blacks and Apollo: Most Couldn't Have Cared Less," *New York Times*, July 27, 1969, E6.

7. Gil Scott-Heron, "Whitey on the Moon," recorded 1970, track 9 on *Small Talk at 125th and Lenox*, Flying Dutchman, vinyl LP.

8. For example, these critiques have been traced in musical traditions of southern black communities. See generally P. J. Blount and Jake Fussell, "Southern Music and Constructing the Space Age," *Journal of Astrosociology* 2 (2017): 121.

9. We should note here that the term "mankind" is itself gendered, and a preferred terminology is "humankind." However, we use "mankind" in discussion of its situated use in the chapter's archive and endeavor to use "humankind" when referring to present policy and law.

10. For the complexity of these long-standing tensions-—and their effect on the historiography of black freedom movements—see Peniel E. Joseph, "The Black Power Movement: A State of the Field," *Journal of American History* 96 (2009): 751–776.

11. On the "benefits of all mankind" language, see chapter 2 in this volume.

12. Tribbe, *No Requiem for the Space Age*, 36.

13. Abernathy, quoted in "Poor People at Launch," *New York Amsterdam News*, July 19, 1969.

14. Williams, memorandum to SCLC Board of Directors.

15. For an extended account of the multiracial coalition aspirations of the Poor People's Campaign, see Gordon Mantler, *Power to the Poor: Black-Brown Coalition and the Fight for Economic Justice, 1960–1974* (Chapel Hill: University of North Carolina Press, 2013).

16. Ronald L. Freeman, *The Mule Train: A Journey of Hope Remembered* (Nashville, TN: Rutledge Hill Press, 1998), 124.

17. Ibid.

18. "Abernathy Asks Antipoverty Liftoff at Apollo Launch," *Boston Globe*, July 17, 1969, 8.

19. "Hit Eastland's Farm in Anti-Hunger March," *Chicago Daily Defender*, July 8, 1969, 8.

20. "Space Head Tells Poor to Unite Behind Apollo," *Boston Globe*, July 16, 1969, 12.

21. "Lift for Poor Urged at Lift-Off."

22. Estimates of the *Community News Service*'s circulation in 1969—during which the paper's distribution was nearing or at its peak—vary, from David Hilliard's 200,000 in his edited collection of the paper, *The Black Panther*, to a US government's estimate—used here—of 140,000. David Hilliard, *The Black Panther* (New York: Atria Books, 2007); Jane Rhodes, "The Black Panther Newspaper: Standard-bearer for Modern Black Nationalism," *Media History* 7, no. 2 (2001): 155.

23. Leigh Raiford, *Imprisoned in a Luminous Glare: Photography and the African American Freedom Struggle* (Chapel Hill: University of North Carolina Press, 2011), 197.

24. In both instances, the Apollo mission is depicted as landing on the Moon as a rocket with a pig-shaped nose, rather than the lunar module.

25. Emory Douglas, illustration, *Black Panther Community News Service*, July 26, 1969, 24.

26. Emory Douglas, illustration, *Black Panther Community News Service*, August 23, 1969, 28.

27. See Sun Ra, *Space Is the Place* (ABC Records 1973).

28. For instance, the July 26 issue—otherwise devoted to speeches delivered at the recent United Front Against Fascism conference in Oakland, California—printed reports on breakfast programs under way in San Diego, Chicago, and Brooklyn. "San Diego Breakfast Moves Ahead," "White Plains Panthers Work for Breakfast," "Panthers' Summer Program," "Brooklyn Panthers Feed Hungry Children," all in *Black Panther Community Service*, July 26, 1969, 15, 20.

29. "Panthers and Young Lords Serve the People," *Black Panther Community News Service*, October 11, 1969, 5.

30. Edward Welsh included in a list of reasons to go to the Moon "the benefits to our standard of living, to education and employment, and to new methods and materials for our productive process will be great even though immeasurable." Shirley Thomas, *Men of Space: Profiles of the Leaders in Space Research, Development, and Exploration*, vol. 7 (Philadelphia: Chilton Books 1965), 258. For Abernathy, these benefits were quite measurable as they were nonexistent within the black community.

31. Frantz Fanon, *The Wretched of the Earth* (1961; repr., New York: Grove Press, 2004), 34.

32. Thomas, *Men of Space*, 245.

33. The distance between rhetoric that serves humankind and the individuals that are considered a part of that population is not new. For example, Sojourner Truth's "Ain't I a Woman" speech plays on a similar theme.

34. Kim McQuaid, "Race, Gender, and Space Exploration: A Chapter in the Social History of the Space Age," *Journal of American Studies* 41, no. 2 (2007): 405–434, at 409–410.

PART II

SOUTHERN CONTEXT

The Newest South

Race and Space on the Dixie Frontier

BRENDA PLUMMER

On October 4, 1957, as the federalized Arkansas National Guard patrolled the halls of Central High School in Little Rock, Arkansas, Soviet engineers and scientists made history by launching the *Sputnik* satellite into space. A new chapter in the Cold War competition between the superpowers had begun. Soviet radio made pointed reference to the United States' racial problems when it informed the world of the moment when *Sputnik* hovered over the troubled American city. Addressing Africa, a continent still largely colonized in 1957, Moscow tried to link its own scientific prowess to African peoples' desires for independence.

Reform-minded Americans also made linkages, attributing US tardiness in space exploration to racism. Talented African Americans, barred from studying physics and engineering at southern universities and consequently from aerospace employment, could not contribute to the country's pool of human capital. In choosing segregation, critics argued, the United States gambled with its own national security. "We spend weeks and weeks and weeks trying to defeat the civil rights legislation," educator Benjamin Mays declared, "while the Russian scientists are moving forward to produce instruments to master the world."[1]

This essay examines the effect of the US space program on race relations in key areas of the South and the impact of that connection to popular culture during an era of shifting racial perspectives, thus linking African American history to the history of technology. It explores the intersection of the struggle for racial equality and space exploration, as both constituted potent narratives of freedom in the American imaginary. Altering the physical environment surrounding the National Aeronautics and Space Administration's (NASA) facilities was central to efforts to probing the heavens. If "terraforming," making uninhabitable planets suitable for human life, was the goal of speculative scientists and science fiction writers alike, it did occur on Earth when technologists and government officials made the natural environment conform to the aims of space research.[2] As in the earlier pursuit of atomic supremacy, the Cold War–driven race to the Moon helped disrupt and transform entire regions. In the process, another version of the New South emerged. The technocratic ideal of progress, coming as it did during an era of both heightened black insurgency and black exodus, elicited political and cultural challenges from African Americans that did not only seek inclusion in the space program but came to question its underlying cultural premises. This essay brings those challenges into conversation with the socially critical lens of science fiction.

Scholars have written extensively about the South's capacity to reinvent itself, including the New South and, more recently, the Sun Belt. Aerospace buttressed the New South's latest incarnation, a region that would finally be absorbed into the national marketplace of culture and consumption. Historically, there had been at least three versions of the New South. The first, articulated in the 1880s by the journalist Henry Grady, looked forward to a modern era of industrialization and urbanization that would leave the cotton plantations and cornfields of slavery behind. Also left behind would be blacks: Grady made it clear that his vision did not include them. A second New South emerged when the Virginia-born segregationist Woodrow Wilson became president and inaugurated reforms in commerce, banking, and agriculture that his successors continued. This was the New South of the Florida land boom, a bonanza that also left the Jim Crow system intact. World War II, the Korean War, and the development of a nuclear arsenal ushered in a third New South of Cold War military bases and atomic installations that continued the pattern of black erasure even as the Pentagon abandoned segregation as a national policy and tried to check it in some communities that hosted bases. Would the fourth New South, the South of rocketry and lunar missions, be different? Initial signs suggested that it would be.

Historians have copiously documented the various New Souths.[3] They have less frequently put them in conversation with one another and with specific social, economic, and political factors that have driven regional change over time. Aerospace endeavors occupy a niche in science history, but the story of NASA was largely written in-house until recently.[4] The physical changes that the space program wrought in the South were the subject of impact studies, but few mentioned a contemporaneous civil rights movement, which authors apparently thought had little relevance to their topic.[5] Such new works as Richard Paul and Steven Moss's *We Could Not Fail: The First African Americans in the Space Program*, Margot Lee Shetterly's *Hidden Figures: The American Dream and the Untold Story of the Black Women Who Helped Win the Space Race*, and De Witt Douglas Kilgore's *Astrofuturism: Science, Race, and Visions of Utopia in Space* are among the first to extensively consider African Americans in context with the space program, and to be placed with conventional publishers. Neil M. Maher's *Apollo in the Age of Aquarius* broadly explores connections to the social movements of the 1960s and 1970s.[6] Most of these works treat the link between aerospace and civil rights as a struggle for black inclusion. The present study is also interested in the interpretive challenges black insurgency posed and how these found creative expression in popular culture.

The space frontier was not the first national project uniting the realm of official policy with strategies for applying science and technology to regional transformation. The earlier pursuit of atomic supremacy had evoked similar possibilities and began the work of physically altering the South to support the new priorities. At the juncture of domestic and foreign policy, responses to the atomic bomb and its links to southern life illustrate the centrality of race relations to the Cold War and US efforts to create a technocratic republic that foreshadowed later aerospace developments.

Federal defense initiatives in the South had provided both a challenge and an opportunity for conventional southern whites. Louisiana parish boss Leander Perez is said to have once asked his governor, Earl Long, "Well, Earl, what are we going to do now that the Yankees have the atomic bomb?"[7] The likely answer, as it turned out, was to share the power. Southern states, with strong histories of military participation in the nation's wars, proved receptive hosts for military and nuclear facilities and benefited from increased government spending and the infrastructural development that national security readiness entailed. Atomic research and weapons production found amenable partnerships with state governments eager to share in the federal bounty. How these changes in local economies and societies would affect African Americans received little forethought.

In South Carolina and Georgia, the Atomic Energy Commission (AEC), in partnership with the DuPont Chemical Corporation, created an enormous nuclear power plant in the Savannah River area, a rural black-majority district, and dispossessed thousands of sharecroppers and farm laborers. Residents were not informed in advance that they would lose their homes. Many resigned themselves to the end of a traditional rural life in the interests of national defense and the hope that the bomb factory would create more regional employment. Some limited consultation with local whites occurred. One enterprising attorney and politician, Strom Thurmond, moved to the area to process claims that whites filed against the federal government. Thurmond handily combined states' rights rhetoric with a subsequent political career of guiding federal dollars to South Carolina.[8]

Black South Carolinians and Georgians experienced blatant discrimination in property assessment and in compensation offers. Once displaced, they were limited to only a few enclaves where they could rent or buy homes at higher prices but lower quality than those available to whites. As during the New Deal and the subsequent postwar era, federal collusion in local patterns of housing bias was nationwide. Black homelessness in the Augusta, Georgia, area increased as residents were pushed aside to make room for incoming defense workers. African Americans found homes in environmentally hazardous areas, near fertilizer plants, pig farms, or railroad tracks. State officials refused to let two black settlements incorporate as towns, thus rendering them ineligible for federal or state aid despite demonstrated need. DuPont and the AEC ignored federal rules that discouraged discrimination by contractors. Black job applicants could only get cleaning jobs at the plant, ensuring that they would remain on the bottom rung of the local economy.[9]

The National Association for the Advancement of Colored People

(NAACP) and the National Urban League attacked bias at the Savannah River Project early in 1951, maintaining that once the project was complete it would be harder to achieve anything other than tokenistic integration. Plant executives and federal authorities had an obligation to the rural people whose lives they had disrupted, NAACP official Clarence Mitchell asserted. "When they take a man off his tractor," he rhymed, "they have to give him a chance to work the reactor." Arguments that Jim Crow could aid America's enemies, however, were then ignored.[10]

Bias at Savannah River was not unique. Mitchell revealed similar problems at the nuclear facility at Oak Ridge, Tennessee. A series of articles about Oak Ridge appeared in the *Baltimore Afro-American* the same year. The plant, built during World War II at a cost of $96 million, boasted a $5 million payroll including skilled workers, but none were black. In the custodial categories where African Americans did find jobs, black janitors received lower wages than whites. As at Savannah River, the housing stock available to black workers was inferior. Segregation was more rigid at Oak Ridge than in some parts of the Deep South. Reporter Rufus Wells "wondered how governmental officials capable of advanced scientific research could be so far behind in their human relations policy." Yet this was a federal project. "All the land and everything on it is owned by the Federal government. No one lives here who is not connected with one of the various plants."[11]

Patterns of discrimination observable at Savannah River and Oak Ridge were replicated in every AEC project. In Hanford, Washington, the federal contractor DuPont faced a labor shortage but would not recruit Mexican workers, claiming such hires necessitated creating another set of amenities, as plant executives thought Mexicans should be separated from others. At the Paducah, Kentucky, gaseous diffusion plant, security guards barred black job seekers from the site, and officials marked applications by race.[12]

The AEC plants changed the physical and social landscapes of the zones where they were created. They dragged rural areas into the domain of the most advanced technology, uprooted entire communities, and enlarged the professional, technical, and managerial classes of the white South. The modernization implicit in atomic research and development in these mushrooming installations, however, did not seem to alter traditional racial thought. Segregation remained a barrier to African American advancement for years to come. Companies like DuPont and Westing-house, in federal partnership, perpetuated through discrimination and anti-union policies a role for the South as a site of low-wage labor. For seg-regationists, the atomic industry's reliance on skilled human capital was an additional advantage: blacks could be excluded in ways that were not overtly racial. Plants still needed African Americans to clean and work jobs that exposed them to radiation, but black aspirations for upward mo-bility in the industry were easily stymied. Black students attended schools made deliberately inferior that narrowed their opportunities. Perceived as part of the rural past, they had no place in the optimistic vision of a future world that scientists, politicians, and local boosters projected.

It was not inevitable that the changes aerospace wrought would repli-cate former patterns of New South development. Race relations were in fact in transition. Yet the structural advantages that the South offered to new industries and technological innovation outweighed reformist im-peratives on the racial front. It is often assumed that NASA, as an instru-ment of modernization, was implicitly allied with civil rights reform and the broad liberalization of American life in the 1960s. While the trans-formation of parts of the Deep South undeniably broke up earlier po-litical, economic, and cultural patterns, aerospace research and develop-ment helped inaugurate a successor regime that neither confronted the structural foundations of inequality nor guarded against the production of new disparities. Unimproved, cheap, and accessible land; proximity to navigable rivers; moderate winters; compliant and conservative politi-cal leadership; and poorly paid workers made growth possible. In 1961, black educator Lawrence D. Reddick speculated that an African Ameri-can youth, "our potentially ideal space man[,] was down there in Mis-sissippi . . . trying to get to the physics books of the public library."[13] The "ideal space man" would have a long wait.

Sputnik had ushered in a feverish race to match the Soviets in space and provided an opportunity to criticize incumbent leaders' presumed un-readiness. Texas senator and majority leader Lyndon B. Johnson, whose ambitions transcended regional identification, found an important role for himself both as a challenger of White House policies and an initiator of change. Senators Stuart Symington, Hubert H. Humphrey, and Estes Kefauver, along with New York governor Averell Harriman, launched rhe-torical attacks on the Eisenhower administration with the assistance of

Adlai Stevenson, presidential candidate in 1952 and 1956. They aimed to score successes for Democrats in the 1958 midterm elections and win the presidency in 1960 with a campaign, according to Stevenson, based on "the loss of our political prestige, Little Rock and our moral prestige."[14]

This approach energized the "scientists, intellectuals, and liberal pundits who had been marginalized as 'eggheads,'" the historian Michael Curtin notes. These elements joined the call for greater military preparedness, better science education, and a rediscovery of national purpose. Steered through Congress by Johnson, the National Aeronautics and Space Act of 1958 paved the way for the establishment of major scientific and industrial infrastructure in the South. Civil rights proponents used *Sputnik* to advocate equalizing opportunity in American society. Ending racial bias, they argued, would greatly enhance the nation's ability to fulfill its democratic commitments and affirm its legitimacy as the leader of the West.[15]

Cooperation from southern Democratic leaders was crucial to the success of any major changes in southern communities. Senator Richard Russell of Georgia opposed the National Aeronautics and Space Act, fearing that it would be used to deny funds to states with segregated public schools. Yet his stance on the federal presence in Dixie, like that of many other southern Democrats, was nuanced. Russell and such colleagues as Senator John C. Stennis of Mississippi, first chair of the new Senate Aeronautical and Space Sciences Committee in 1959, and John J. Sparkman of Alabama, a major proponent of missile development, had supported New Deal efforts to bring economic development to a region that President Franklin D. Roosevelt had called the nation's number one economic problem. "States' rights" had not been the rallying cry when it came to rural electrification and constructing army bases.[16]

The momentum *Sputnik* achieved intensified when, in 1961, cosmonaut Yuri Gagarin made the first manned space flight. Soviet propagandists reprised the theme of solidarity with the colonized world with a poster proclaiming a "New Day" that depicted a black man bursting from his chains and saluting Gagarin as the Soviet pilot flew over Africa. Soviet geophysicist Yevgeni Fyodorov noted, "Comrade Gagarin saw the Congo where only recently Lumumba, the valiant champion of the happiness of the Congolese people, was heinously murdered." Fyodorov and other Soviet celebrants linked the USSR's accomplishment to a critique of Western imperialist exploitation of vulnerable countries and territories. The year before, when seventeen African countries became independent, the USSR established an Institute of African Studies within its Academy of Sciences.[17]

Gagarin's flight accompanied swift political and social changes in the United States. As *Air & Space Magazine* observed, "Gagarin's Earth orbit, the failed Bay of Pigs invasion of Cuba, Alan Shepard's flight, the Freedom Rides with their attendant violence and imposition of martial law, and Kennedy's man-on-the-Moon-by-the-end-of-the-decade speech all happened within weeks of one another in 1961." US officials were aware of the USSR's determination to capitalize on the turn of events. "In seizing an early lead in space and following it with a series of dramatic successes," the Central Intelligence Agency commented, "the Soviets have sought to bolster, both at home and abroad, claims of the superiority of their system."[18]

American leaders were thus determined to match the Soviets point by point. Just as the Soviets recruited their best scientists and technicians to the space effort, so did the Americans. In a move that recalled Gagarin's exploit, astronaut Gordon Cooper Jr. orbited over Africa during his nearly day-and-a-half-long Project Mercury flight in 1963. He radioed salutations to the nascent Organization of African Unity.[19] This was not simply a conventional diplomatic gesture. The United States found it easier to match Soviet gestures than to tackle the domestic race question head on. Both the Eisenhower and Kennedy administrations were reluctant to fully invest in restoring the civil rights lost to blacks in the collapse of Reconstruction. Kennedy focused instead on Africa in a bid to mollify increasingly critical black voters. He could thereby nod to their concerns while avoiding a confrontation with segregationists. The approach also aimed at neutralizing for Africans the appeal of Soviet anti-imperialist rhetoric. It underlay White House invitations extended to African leaders and visits to African countries by astronauts Cooper and Pete Conrad.[20]

Camelot-era symbolism belied, however, the very regional character of the space program and its associated economy. NASA built major facilities in southern states made attractive by favorable weather, prior histories of military base development, and a compliant population. Lyndon Johnson, first as a senator and later as vice president and president, worked hard to promote the agency's mission as devoted to civilian control of aerospace research. Therein lay a benefit. As the historian Joseph A. Fry put it, "the space program went far toward fulfilling LBJ's search for a mechanism for building a New South of 'science and technologically-based enterprise.'" Beyond the conventional pork barrel associated with federal largesse was something more far-reaching: a fresh iteration of the New South.[21]

Space and its management in a terrestrial sense played a part in the society NASA helped to produce. "The space age has arrived on the Gulf Coasts of Mississippi and Louisiana," the *New York Times* proclaimed, "displacing people, stills, snakes, alligators, wild pigs, ducks, a graveyard and attitudes." Scientists and engineers terraformed vast acreage to construct the Marshall Space Flight Center (MSFC) in Huntsville, Alabama; the Michoud Assembly Facility near New Orleans; the Mississippi Test Facility at Pearl River (now the John C. Stennis Space Center); the Launch Operations Center at Cape Canaveral, Florida (later the Kennedy Space Center); and the Manned Spacecraft Center in Houston. These sites helped create the so-called Sun Belt, where the size and wealth of a technocratic, professional white middle class came close to equality with the national average, exceeding it in some areas. Educated white newcomers to southern communities encountered revised patterns of racial division.[22] "Well-guarded out-of-reach property values insured the perpetuation of a homogeneous society of middle- and upper-middle-class families," Paul Gaston writes, "thus helping to insure a new form of segregation. As this separation was accelerating, class now joined or even supplanted race as the primary dividing line."[23]

The emerging paradigm comfortably fit the highly adaptable German scientist Wernher von Braun. Von Braun, MSFC director from 1960 to 1970, and the public face of NASA's scientific expertise, was brought to the United States in 1945 to conduct rocket research for military purposes. His Nazi party membership and cold-blooded use of slave labor at the horrific Mittelbau Dora concentration camp was quietly forgotten, as was his role in the development of the V-2 rocket that wreaked havoc on London during World War II. Once in the United States, von Braun became a born-again Christian and a noted presence at MSFC prayer breakfasts. This religious conversion helped him fit in handily among southern evangelicals as well as in larger US aerospace and defense circles. While his rocket expertise is commonly known, his role as a progenitor of the new southern elite is not acknowledged.[24]

Von Braun performed a third role as a purveyor of the imaginary that sustained NASA's popularity. Humans had a divine mission, he declared. "Only man," von Braun wrote,

> was burdened with being an image of God cast into the form of an animal. . . . And only man has been bestowed with a soul which enables him to cope with the eternal. . . . If man is Alpha and Omega, then it is profoundly important for religious reasons that he travel to other worlds, other galaxies; for it may be Man's destiny to assure immortality, not only of his race but even of the Life spark itself. . . . By the grace of God, we shall in this century successfully send man through space to the moon and to other planets on the first leg of his last and greatest journey.[25]

There was a note of sadness in this apotheosis. Man's greatest achievement would be to leave the planet rather than to embrace it. As a prophet of the future who had experimented with science fiction writing, von Braun portrayed the Space Age as one of "eternally renewable freedom" or, more darkly, as "escape."[26] His vision would color the imaginative thinking of scientists and infuse popular culture for years to come.

Historians have argued convincingly that acquiring information was a secondary priority for space funding. Presidents Eisenhower, Kennedy, Johnson, and Nixon had all expressed some skepticism about space exploration's scientific worth. Eisenhower complained of the cost. They yielded, however, to the realization that, in addition to the prestige competition with the Soviets, the space program would do important cultural work at home by holding out the possibility that age-old science fiction dreams might be fulfilled. Heroic astronauts would reaffirm the values of patriotism, self-sacrifice, bravery, and adventurous masculinity, traits threatened with disappearance during an age of presumed effeteness. Candidate John F. Kennedy had won the 1960 presidential election in part by impressing voters with his claim that America had gone soft and needed to regain the hardiness of its frontier heritage.[27]

These cultural imperatives were powerful incentives for the rapid development of the Sun Belt and the aerospace industry. The ideology of progress embedded in Space Age development offered a carrot and a stick for integrationists and segregations alike. The carrot led proponents of integration to view the Space Age as a pathway to a racially just society, while the stick threatened to bar the way through inventive methods of segregation. Segregationists, enticed by the carrot of federal support, came to understand that government resources might be tied to discipline. Vice President Johnson linked progress in aerospace to "revolutions . . . in our industries, in our systems of education, in our hiring policies, in the realms of science, law, medicine, and journalism." "Because the Space Age is here," he informed participants at a Seattle conference, "we are recruiting the best talent regardless of race or religion, and, importantly, senseless patterns of discrimination in employment are being broken up."[28] Houston, where fortunes were quickly made in aerospace, provided an example. When the local utility company refused electricity to a naval base to protest the navy's insistence on an antidiscrimination clause in its contract, LBJ informed the company that Houston could lose millions in federal contracts for the NASA satellite tracking station if bias persisted. The utility capitulated.[29]

Rice University in Houston was another target in the sights of modernizers. Rice had benefited greatly from the Manned Spacecraft Center's arrival. Large sums of federal money, private contracts, and scientists transformed the university as it developed the only space science department in the country. Yet Rice in 1963 still barred black students, a prohibition the founder had specifically mandated. Rice's own trustees went to court seeking to overturn this Jim Crow legacy, understanding that it threatened the new largesse. "Rice University today stands at the crossroads," its attorney argued. "It could go to the moon or it could return to the 19th century." Rice admitted its first black student in 1965.[30]

Lip service to civil rights laws could be rewarded by federal munificence, while blatant resistance might be punished. One southern white man expressed this sentiment to a journalist, saying, "I'm like most Southerners when it comes to passing judgment on the civil rights bill. That is, I don't like it, and I guess I'm lined up with Goldwater on that point. But if it means I've got to have a choice of letting colored people sit next to me in a restaurant or having Cape Kennedy closed up, well brother, you can bet I'll let them sit. Not only that—I'll pass them the sugar." The bargain that this man and others were willing to make enlarged southern whites' stake in militarism, bound them more closely to the national government and its priorities, and strengthened the appeal of a racial politics revised, but not abandoned, for modern times.[31]

Events in Brevard County, Florida, home of the Kennedy Space Center, further illustrate the ways that tradition was challenged yet found breathing space within the new order. The county was changing from the backwater where the writer Zora Neale Hurston had waited out her final years in poverty, and where the Ku Klux Klan on Christmas Eve 1951 had assassinated NAACP official Harry Moore and his wife, Harriette, in the tiny town of Mims. It was now home to a resort that bathed in the reflected glory of the astronauts. These men favored the appropriately named Starlite Motel in Cocoa Beach as their preferred temporary residence. In October 1962, the Starlite hired the popular black singer Della Reese to perform, but there was a hitch. Unlike its companion city Cocoa, Cocoa Beach was a "sundown town" where African Americans could not stay overnight. Reese additionally would have to appear before segregated audiences. The motel arranged alternate performance days for the different races but backed down in the face of bomb threats from racists. Starlite's managers then arranged for Reese to sing before African Americans at the local black high school in Cocoa.[32]

Once Reese arrived at the school, 2,000 youthful protesters, including members of the Brevard County NAACP Youth Council, confronted her for accepting Starlite's terms. She then refused to perform before *any* segregated audience. Starlite capitulated and agreed to host integrated audiences. Reese's concert went off without a hitch, and, as a by-product, the nearby erstwhile segregated Cape Colony Inn, partly owned by astronauts, hastened to accept black diners. African American attendees loitered in Cocoa Beach, defying the ordinance that proclaimed they had to leave town by 11 p.m. Reese spent the week of her tour residing at the Starlite, the first black guest to do so.[33]

Ironies abounded in the incident. While Cocoa Beach was a sundown town, its first non–American Indian residents had been African Americans freed after the Civil War. *Pittsburgh Courier* columnist Izzy Rowe commented on the rapid disappearance of the sundown statute in a city where "all the major missile contractors by law are equal opportunity employers." He noted that "the State Dept. has brought Afri-Asian VIPS to the Cape area on more than a few occasions," thus simultaneously blurring the distinct racial lines created by segregation and bringing the bias against African Americans into sharper relief.[34]

The seeming disappearance of formal Jim Crow in Brevard County was partial. Brevard Junior College, the only institution of higher learning in the area offering technical and scientific courses, did not admit black students as late as 1962. The African American population relative

to whites declined as new aerospace facilities displaced black neighborhoods. Uprooted residents were sent to a swampy unincorporated area,
infested with snakes, where they could not vote in city elections. Some
had owned solid houses and were judged too affluent to qualify for public
housing. A local corporation, Canaveral Groves, Inc., advertised land for
sale in these outlying areas in *Ebony* magazine. For NAACP secretary
Roy Wilkins, the new aerospace community had risen "on foundations
of racial segregation and discrimination with federal assistance." More
than a half-century later Cocoa and Cocoa Beach remain overwhelmingly
white.[35]

The historian Walter McDougall asserts that "technological infusion
was to call to life a New South" whose emerging order represented "the
economic and intellectual component of the Second Reconstruction."[36]
Historians use that term to refer to the transformations accompanying
the civil rights movement's legislative successes. I suggest instead that
the effect of technological change was to sidestep a Second Reconstruction. Astronauts' presence as local entrepreneurs and popular figures lent
prestige and visibility to Cocoa Beach and underscores the federal role
in underwriting the demise of formal Jim Crow and inaugurating a successor regime of structural segregation. What made it different from earlier patterns that paired southern growth with racial exclusion is that it
occurred at a historical moment in which black demands could not be
denied. While it was no longer possible to legally bar African Americans
from southern towns, popular projects based on technological skills often absent among black southerners could effect an end-run around civil
rights goals. "Federal programs," write Matthew Lassiter and Kevin Kruse,
"played the most critical role in building the suburban South and simultaneously standardizing patterns of residential development and housing
segregation across the nation. The Federal Housing Administration subsidized the private construction of all-white subdivisions and permitted
open racial discrimination in the sale and rental of federally supported
housing well into the 1960s."[37] Roy Wilkins complained about housing
bias, noting the approximately 300 vacant Federal Housing Authority–
financed houses in the Cocoa area. In 1960, Brevard County minorities
numbered 11 percent of the population. A 1964 report for the Committee
on Space of the American Academy of Arts and Sciences noted that "as
farm land is taken up for the space center and the support contractors,

Negroes may actually be displaced from their jobs as farm laborers and move elsewhere in search of work. The size of the Negro community remains the same while that of the white community multiplies." By 1966, blacks constituted only 8 percent of Brevard County residents.[38]

Under the circumstances, extending formal citizenship rights to blacks was not only less troubling but a vital aspect of this process. Those rights need not mean the violation of white spaces. The civic—not social—incorporation of blacks could strengthen the nation. Diane Nash, leader of the historic Nashville sit-ins and a founder of the Student Nonviolent Coordinating Committee (SNCC), had "connected the student movement with the struggle against communism." If blacks had an equal chance for educational opportunities, she argued, "maybe some day a Negro will invent one of our missiles." If moral imperatives did not constitute sufficient arguments for granting civil rights, retrieving wasted human capital might be more convincing. The human capital argument fit in nicely with the era's new science of development, especially as applied to emerging countries and "backward" peoples. Yet by the end of the decade, black criticism of the space program would not rely only on demands for inclusion.[39]

Many regional conservatives abandoned the formal rituals of Jim Crow but continued to oppose the social investment that would raise the status of the least privileged part of the population. All but the most obdurate white supremacists were beginning to realize that the "white" and "colored" signs in public places were superfluous. The regime that necessitated rigid control of black people underwent a process of modification, setting the stage for the South's reincarnation as part of the Sun Belt configuration. *Life* magazine helped inaugurate the phenomenon in a lavish February 2, 1962, photo spread titled "The Spell of Southern Warmth: *Life*'s Gulf Coast and Florida Tour." Readers shivering in the bleakness of a northern February could gaze wistfully at vivid photographs of beaches, plantation manors, colorful foliage, festivals, and beauty queens without seeing a single nonwhite face. Dixie's new identity would bury its violent past, with which African Americans were ineradicably associated.[40] This eradication was central to the narrative of progress and the implicit reunification of the country, as the South would no longer be marked as peculiar. Purged of backward people, it could take its rightful place in an expanded national culture. Though protest continued, scenarios based on race-less technological success provided little traction for civil rights–based objections.

Sporadic resistance of the old-fashioned kind did continue, as in Louisiana, where as late as 1964 the state rejected federal funds to support job retraining even as NASA was having difficulty finding local workers sufficiently qualified for employment at the Michoud assembly plant, the vast installation named for the antebellum plantation that once occupied part of its site. Diehards apparently disliked the requirement that training be integrated and so reconciled themselves to higher local unemployment rates.[41] They thus perpetuated a feedback loop where prejudice impeded the growth of the economy.

The shortage of black engineers and technicians reflected bias, generations of educational deficits, and the reluctance of skilled African Americans—as well as many whites—to relocate to persistently rural southern areas. Development remained uneven as urban blight and depopulating farming areas continued to characterize the region even as new white suburbs were being constructed. Committed segregationists gave up gradually and reluctantly. Alabama made international news in the early 1960s as its devotion to the old order led Governor George Wallace to physically block two African American students' entry to the state university, and when photographs of police dogs attacking demonstrators in Birmingham flashed across the globe.

Huntsville, Alabama, home of the MSFC and of Alabama A&M College, where writer Alex Haley grew up as a faculty brat, was also a segregated town. Wallace was unpopular, however, among Huntsville's educated elite. When he visited the MSFC in 1964, von Braun asked him if he would like to go to the Moon. "Better not," Wallace replied, "you fellows might not bring me back." Formal barriers to equality in public accommodations in the Rocket City fell relatively quickly because of federal benefits and pressure. Huntsville, Lady Bird Johnson told her husband, was "such a bustling center, and seems to be really tying the South to the future."[42] Yet historically black Alabama A&M College, now University, also in Huntsville, did not figure in this. According to rocket scientist Ernst Stuhlinger, race was not the issue. Instead, Alabama A&M president Richard D. Morrison appeared to be tied to the past, rejecting advances made by NASA because the black school was meant to create better farmers, not astronauts.[43]

In a December 1964 speech to the local Chamber of Commerce Wernher von Braun called for moderation on racial matters and conformity to the Civil Rights Act of 1964. After widely disseminated film footage recorded Alabama state police whipping and bludgeoning demonstrators on a bridge in Selma in 1965, the former Nazi endorsed the Voting Rights Act.[44] The rural dispersal of whites to cities and suburbs, a process abetted by the aerospace industry and other federally subsidized initiatives, helped promote the streamlined conservatism that replaced the old racist politics. Economic restructuring minimized intergroup friction by prompting a black exodus, leaving in (separate) place mostly those too poor to go. Civil rights struggles at polling stations, courthouses, and schools barely touched the charmed existences of those who could eliminate race—and blacks—from their social and spatial reality. There would be no negative international press if there were no visible victims.[45]

Yet civil rights insurgency continued to challenge the wholesome tale of progress. When, during the 1964 World's Fair, demonstrators from the Congress of Racial Equality (CORE) denied the relevance to African Americans of the "Lunar Fountain," the "Fountain of the Planets," and the "Unisphere," CORE director James Farmer explained their intent. The protest was "designed to point up the contrast of the glittering fantasy world of technical abundance and the real world of discrimination, poverty, and brutality faced by the Negroes of America, North and South." "For every piece of bright chrome that is on display," he asserted, "we will show the charred remains of an Alabama church. And for the grand and great steel Unisphere, we'll submit our bodies from all over the country as witnesses against the Northern ghetto and Southern brutality."[46] Civil rights organizations also condemned NASA's tracking station in South Africa, where the US government acquiesced in apartheid, sending no African Americans to work there and agreeing to limit the shore leave of black troops.[47]

While conventional civil rights leaders promoted orderly demonstrations, discontent could not always be controlled. State and local police, the National Guard, and the US Army were variously deployed in American cities during the "riots" of the 1960s. Token efforts at economic improvement followed repression. The Aerojet Corporation, best known for rocket engine design and subsequent contributions to the Moon launch, responded to the government call to increase black employment in Los Angeles after the 1965 Watts rebellion, opening a plant in Watts that made tents for the military.[48]

A tidal wave of revolts occurred in the spring of 1968 following the assassination of Martin Luther King Jr. In response, the Pentagon created the Joint Chiefs' Contingency Fund and planned a secure facility from which it could guide military operations against homegrown insurrectionists. Keeping African Americans on the margins of society denied their centrality to the nation, but violent repression suggested that containing black rebellion was fundamental to national security. Federal authorities coordinated their activities with firms that pioneered sophisticated technology. Aerojet, no longer confining itself to ghetto tent-making, developed a tank for use in quelling urban revolts. The virtually impregnable vehicle, equipped with a night vision camera, was impervious to improvised explosive devices, and its interior, in the company's words, boasted "the aesthetics and comfort features of a luxury passenger car."[49]

In Houston, astronaut Buzz Aldrin, a veteran of the 1966 Gemini 12 spacewalk and slated to be a member of the Apollo 11 mission to the Moon, attended an April march commemorating King. "A few days after Dr. King's death," he later recalled, "I called my pastor in Houston to join me in a 'walk'—as we participated in a memorial march through the streets of downtown Houston in honor of Dr. King's life and all he'd fought for in the civil rights movement." The novelist Norman Mailer applauded his initiative. "In the political straits of NASA it was a brave act," Mailer commented. "It could have affected his career." King's antiwar position had not endeared him to hawks, and the astronaut's sympathy for the martyred civil right leader could cast doubt on his fitness for the Moon. "Little could damage a man more than to be considered sympathetic to peace in Vietnam while working in Houston," Mailer observed.[50]

These responses to the aftermath of King's assassination signaled the complex character of a changing society. Aerojet, combining roles as space purveyor, employer of last resort, and hired gun, and Aldrin as protagonist in two dramas, one, the relentless determination of straight-arrow white males to conquer space and the other an expression of terrestrial empathy and consolation, expose the instability of the moment. Charity accompanied repression and largesse challenged indifference against the backdrop of black revolt and the broad imperative to rethink the racial order.

Seeming worlds away from America's contentious racial politics, scientists and engineers planted the Stars and Stripes on the Moon in July 1969. Television sets everywhere locked onto the inky lunar night as white Americans, clad in white, touched down gently and bounded lightly across the eerie sands of a satellite that, from Earth, appeared flawlessly white.

The Moonwalk concealed black discontent amid national celebration. NASA had already begun to diversify the astronaut corps when air force pilot Robert H. Lawrence Jr. was selected for training in 1967. Officialdom was attuned to how the selection of a black astronaut would be interpreted. "I suppose its international implications are even more important than the domestic political one," an attorney once wrote to President Kennedy's deputy special counsel. "The Russians don't have any Negroes to shoot into space." The Russians solved this problem later by bringing the Afro-Cuban pilot Arnaldo Tamayo Méndez into their cosmonaut program, beating the United States in this endeavor by three years.[51]

Aside from the initial dearth of black astronauts, conditions on the ground at the Kennedy Space Center where the Apollo 11 launch took place blighted this pursuit of prestige. Brevard County, Florida, had erased the outward manifestations of Jim Crow by 1969 but continued to face Earth-bound problems of hunger and malnutrition. "Even as a million converged on Cape Kennedy to witness the start of Apollo M's epic voyage, thousands of Brevard County inhabitants who qualified for federal food aid were being denied because the local authorities had not set aside matching funds," Dale Carter observes. "As the only Black doctor in the locality remarked, 'I guess they're more concerned about promoting tourism in Brevard County than caring about hungry people.'"[52]

The Southern Christian Leadership Conference (SCLC) strove to highlight the incongruities, bringing low-income families from five southern states to Cape Kennedy in time for the July Moon launch to highlight the problems of hunger and unemployment throughout the country. SCLC leader Ralph Abernathy claimed to represent "the people of the 51st state of hunger." Some of those accompanying him held signs with such messages as "Rockets or Rickets?" and "Billions for Space and Pennies for Hunger." "To the nation's black poor," *Ebony* magazine editorialized, "watching on unpaid for television sets in shacks and slums, the countdowns, the blastoffs, the orbitings and landings had other-worldly alienness—though not the drama—of a science fiction movie. From Harlem to Watts, the first Moon landing in July of last year was viewed cynically as one small step for 'The Man,' and probably a giant step in the wrong direction for mankind."[53]

Yet the glamour of the Moon shot muted the thrust of SCLC's protest. Mules, the staple work animal for generations of black tenant farmers and sharecroppers—and hardly Space Age transport—accompanied 300 demonstrators who arrived at the launch site. NASA director Tom Paine cordially greeted Abernathy, asked the black minister to pray for the astronauts, and invited some one hundred protesters to view the launch in the special guest area.[54]

Moderate civil rights advocates were not alone in criticizing the space program. The radical *Black Panther* newspaper anticipated the Moonwalk in the January 25, 1969, issue, and featured a cartoon titled "Imperialist Plans." The drawing depicted otherworldly scenery where police, portrayed in Panther fashion as pigs, drove a group of slaves. One pig orders: "Hey handle those slaves with care we're gonna need them for Mars, Pluto, and all the other planets." To Black Panther member Connie Matthews, the Moon landing powerfully exemplified the global reach of imperialism. "We must stop talking in terms of countries," she declared, "we have to talk about internationalism because the United States has now gone to the moon, they will go to Mars, they will go to Venus next, so that it is not just a question anymore of planet earth."[55] Panther artistry's bleak lunar landscape inscribed science fiction in the present moment: reading the genre's focus on exploratory conquests as a real-world fait accompli. The Moonwalk that delighted mainstream America was a looming disaster for these detractors. "Visions of catastrophe appear in large part to be the symmetrical opposites of colonial ideology's fantasies of appropriation," writes John Rieder.[56] In the absence, furthermore, of active black subjects in conventional science fiction imagining, the future portended slavery for African Americans.

Concern was expressed in other quarters. In the wake of the foreign and domestic crises of the 1970s, Congress and the White House became less interested in supporting space ventures. Representative Ed Koch, later mayor of New York City, declared, "I just for the life of me can't see voting for monies to find out whether or not there is some microbe on Mars, when in fact I know there are rats in the Harlem apartments."[57] The radical African American musician Gil Scott-Heron also saw the Moon shot as a distraction. His 1970 song "Whitey on the Moon" focused on its irrelevance to the struggles of black working people. On launch day, July 16, 50,000 black New Yorkers were not glued to their television screens. Instead, they attended a soul music festival where "the single mention of the LM [lunar module] touching down on the moon brought boos from the audience." Similar reactions were reported nationwide.[58]

SCLC's protests continued into 1971, when the organization sent 200 demonstrators to the Kennedy Space Center on a "march against moon rocks." The occasion was the launch of Apollo 14. March participants included a group of black chambermaids from Daytona Beach motels forced to work for $35 a week when the federal minimum wage would have netted them $64. March leader Hosea Williams explained, "We are not protesting America's achievements in space, we are protesting our country's inability to choose humane priorities." Williams found Apollo 14 frivolous: he accused the government of spending $3 billion to collect rock samples that would be used solely to distribute prestigious souvenirs to foreign leaders. This time, NASA accorded demonstrators less than special treatment, sending them to watch the launch from a site "a good distance from the VIP site where they had expected to be."[59]

Many commentators focused on America's technological prowess and its contrasting backward race relations, calling for stronger efforts to include blacks in aerospace endeavors. Yet the space program also evoked imaginative responses that sprang from literature, music, and popular media. It was difficult not to notice the compelling visual imagery of outer space. Astronaut Frank Borman wrote of the peculiar aesthetics of extraterrestrial travel as he and two companions left Earth's orbit during the earlier Apollo 8 mission. "I felt a strange sense of detachment up there," he recalled, "as if this nine-by-thirteen-foot box had suddenly become a tiny, self-contained world of its own." It was "as if our entire existence had been compressed into an environment of winking amber and red instrument lights." Bright light, glass, and shiny metallic materials contrasted sharply with the amorphous and illegible blackness beyond the spacecraft.[60]

Left unsignified, or perhaps part of that darkness, were the Earthbound, unskilled, unqualified masses of color. Their structured absence in early science fiction—as well as in high-level NASA jobs—suggested that they would never experience interstellar travel. Like those left behind in dilapidated cities when whites headed for clean and sunny suburbs, they seemed doomed to future abandonment when the anointed departed a polluted and irretrievable planet for the stars, worlds where people of color had no place. As recently as 2014 many sci-fi fans protested when the latest film in the *Star Wars* series featured a "stormtrooper" of clearly African descent. "Blackness gets constructed as always oppositional to technologically driven chronicles of progress," writes Afrofuturist scholar Alondra Nelson.[61] Sensitive to the ways in which worldly understandings of hierarchy affected the interstellar imaginary in both science *and* science

fiction, many African Americans perceived the future of space exploration as one that would mirror the historical past. As in the so-called Age
of Discovery, when Christopher Columbus and others harnessed the in-
digenous populations of Africa and America to their projects, outer space
would be a place of servitude and subjection.[62] Yet many black detractors
proved unwilling to dismiss the power of the imaginary and sought to
bend it to their own purposes of critique.

Science fiction addresses the social and political problems that society
finds difficult to solve or even discuss. Its themes rehearse and mull these,
proving a way to displace cultural and ideological dilemmas. "Science fiction is actually transmitting assumptions of racism even in stories that are
ostensibly envisioning a future where race has become irrelevant," Isiah
Lavender asserts. This is exemplified in the popular science fiction television show Star Trek, which created controversy in November 1968 when
Lieutenant Uhura, played by African American actor Nichelle Nichols,
and Captain James T. Kirk, played by white actor William Shatner, kissed.
This was the first interracial kiss broadcast on syndicated US television.
It could not be portrayed as spontaneous or natural. The story line described the kiss as the result of mind control exerted on Kirk and Uhura
by hostile extraterrestrials. Only in the environment of outer space could
such a breach of propriety be imagined, and only as a television foray into
the fictitious could it be performed before the public. NASA, in an ironic
conflation of the imaginary and the real, later employed Nichols to recruit
young people interested in science.[63]

African Americans had two options regarding the respective institutional racism of NASA and the cultural racism of popular science fiction.
The first, pursued in tandem with the government, was the integration of
the astronaut corps and abolition of job bias in aerospace-related industries. This should finally settle the question, at least on the symbolic level,
of African Americans' relationship to the US space venture. NASA would
thus lead in opening jobs across an array of technical and nontechnical
specialties to people of color. Instead, it earned a reputation as one of the
most recalcitrant federal agencies when it came to both racial and gender
desegregation.[64]

As for women, author Kim McQuaid notes, "the only three females
NASA had so far sent into space [by 1973] were two spiders and a mon-
key." The complex nature of space research and astronaut training provided the agency with an excuse to maintain its exclusion policies, but not

every aerospace job required an advanced physics or engineering degree. Indeed, during World War II NASA's predecessor, the National Advisory Committee for Aeronautics (NACA), hired women to do calculations so that engineers could otherwise use their time. Though skilled and college-trained, these female "computers," as they were called, ranked as "subprofessionals" and received below market-rate salaries. This practice continued when NACA became NASA in 1958. The computers included a few black women who worked in segregated facilities until NASA abandoned Jim Crow in the 1960s. A half-century later, they were finally recognized in the Hollywood film *Hidden Figures* and by the awarding in 2015 of the Presidential Medal of Freedom to former NASA employee Katherine Johnson.[65]

Ironically, the highest-ranking woman in the agency was black, Ruth Bates Harris. NASA hired Harris as equal opportunity director in 1971 but demoted her before she had even begun the job. If NASA could claim with some justification that it found women and minority engineers and scientists hard to locate, it could not make this argument for the many other job classifications that it had managed to avoid integrating. Harris blew the whistle on this foot-dragging and was subsequently fired.[66]

The project of racial integration was not the only black response to the space program's challenges. The second option lay in the capacity of popular culture not only to parody the space program but also to reinterpret its meaning and offer alternative futures to the ones outlined by the technocratic state. In this regard, astronauts themselves provided models by merging their practical business acumen with the public's hunger for make-believe and success. John Glenn, the first American to orbit Earth, as early as 1959 had joined other astronauts in investing in Florida land and hotels, further advancing the Sun Belt takeoff. His fortune helped bankroll his successful 1974 run for the Senate in Ohio and his 1984 presidential campaign. Glenn and other Mercury colleagues had a minority interest in Cocoa Beach's Cape Colony Inn where, unbeknownst to them, the management had refused lodging and restaurant service to a black couple from Glenn's home state. Another of Glenn's hotels was near Walt Disney World, where the Epcot pavilion would celebrate the promise of future space forays. Disney World was based on Walt Disney's earlier California project, Disneyland, where Tomorrowland was already whetting the appetite for astronaut heroes and extraterrestrial adventure.[67]

Disney's entertainment empire fully exploited public curiosity and fascination with the stars. It enlisted von Braun's services to promote *Disneyland*, its popular and enormously influential television program. The rocket scientist worked diligently with Disney to translate public enthusiasm into political pressure for continued NASA funding. "Contrary to broadly received opinion, scientific thought is not transcendent; it exists within historical and cultural matrices," Cedric Robinson asserted. "The most powerful economic, political, and cultural impulses of a social structure impose themselves as codes and desires on the conduct, organization, and imagination of scientists."[68] By the time that outer space themes began to figure prominently in African American popular culture, tightly knit mainstream networks linking business, science, and entertainment had long been promoting various space-related enterprises and relying heavily on fantasy constructions to stoke imaginative fire as well as to sell products.[69]

Budgeting for NASA had begun to decline after 1967, even before Apollo 11 captured global attention. The 1970s was a period of further retrenchment for the space agency. An economy that had long assumed its infallibility found itself distressed. Several factors burst the bubble: the costs of the Vietnam War, the collapse of the Bretton Woods international monetary system, the Organization of Petroleum Exporting Countries' embargo and the sharp increase in the price of energy, and competition from more efficient foreign industries. Planners faced stagflation, a combination of inflation coupled with high rates of unemployment in heavy industry and construction. As budgets shrank, many NASA specialists, including von Braun, transitioned to private-sector employment.[70] African Americans, historically at the bottom of the economic pecking order, suffered double-digit official unemployment rates. In an atmosphere that forced the postponement of celestial journeys, black critiques restaged the lost aspirations of politicians and engineers.

Inclusion in a fading space program was not the goal. Restaging took place in the imaginative realm. While our observations about science fiction are constrained by the formalities of treating it as the province of the unreal, the genre mediates the society's essential concerns. "Speculative fiction is as saturated with race thinking as any other variety of popular culture, and it tends to reproduce conventional understandings of race," Andre Carrington argues. Just as conventional science fiction had always focused on disjuncture and incongruity, works by African American science fiction authors referenced the past and explored how, in Lisa Yaszek's words, "African slaves and their descendants experienced conditions of homelessness, alienation, and dislocation that anticipate what philosophers like Nietzsche describe as the founding conditions of modernity." Science had helped to create the conditions of an African American displacement that found cultural expression in science fiction's tales of exception. "The alien encounters and interplanetary abductions people experienced as delusions in the Cold War present," Kodwo Eshun writes, "had already occurred in the past, for real."[71]

Works by such writers as Octavia Butler and Samuel Delany centered on alienated protagonists and themes that reflected on slavery, loss, dystopia, and survival. Through the heroines of Butler's sensibility, for example, community persists despite urban crises, nation-state collapse, and the disintegration itself of what it means to be biologically human. Delany's take on slavery, transmogrification, and the erotics of power maps aspects of the African American past as much as it imagines an alien future in which terrestrial commitments have not been entirely laid to rest.[72]

The 1970s funk music group Parliament Funkadelic played a variation on the theme. Critic Adilifu Nama describes how its aluminum foil–clad astronauts "assail[ed] the privileged place that white heterosexual male authority holds . . . frequently clothed in the iconography of American militarism and scientific intellectualism. All of the typical symbols of white authority and institutional power—scientists . . . generals, army heroes, and even the president—are thoroughly lampooned."[73] Appearing in costumes that reflected the production values of such relics as the original *Dr. Who* series, Parliament Funkadelic's hokey space travelers toyed with the conventions of space flight and science fiction. The Smithsonian Institution recently recognized the funk group's influence by acquiring the Mothership for the recently opened National Museum of African American History and Culture in Washington, DC, where the august spacecraft will be part of a permanent installation. Just as the Smithsonian museums became the final resting places for the Apollo command modules when NASA could no longer afford to maintain them, the Mothership, Apollo's referent, will similarly be a venerated cultural object.[74] Ironically, the apocalyptic fantasies entertained by Parliament Funkadelic were not far removed from those of Wernher von Braun, who invoked end-times in his vision of man's forays into the void.

Jazz also reflected an awareness of the possibilities of space. No musician developed this more than Sun Ra, whose ruminations about ancient Egypt fused with his apocalyptic visions of future extraterrestrial life. In Sun Ra's concerts and his 1974 film, *Space Is the Place*, he refused to accept the terms in which he was conventionally defined and claimed to be an extraterrestrial sent to convey a message of exodus. At the film's end, a select group of people of color are transported into the ether, leaving Earth to its fiery destruction. Sun Ra's choice to conflate the extraterrestrial with ancient Egypt functions as both a back-to-Africa narrative and defiance of Western time, with its implicit teleological assumptions. His performances flouted the futuristic bent of mainstream science by reasserting the significance of an ancient past with its own arts and sciences, and by invoking the millennial tradition in African American thought.[75] This pastiche suggested a postmodern frame for engagement with space adventures in the black imaginary: if modernism *produced* America's space program, then postmodernism was a way to *reproduce* it.

In the early twenty-first century, the modernist dream of space flight has been revived by wealthy entrepreneurs planning to bring extraterrestrial travel into the neoliberal economic sphere and loosen the ties that have so far bound it to governmental exigencies. These magnates dream of creating an "Earth-to-Mars economy." In a 2008 article titled "Billionaires in Space," the business magazine *Forbes* quoted $200,000 as the price of the ticket to be a "private astronaut."[76] Apparently, 99 percent of the world's population will *not* be cleared for takeoff.

This exclusivity mirrors the marginalization of the past, as disparate resources and long-standing biases continue to condition the search for extraplanetary freedoms. On Earth, the ongoing evictions and displacements of the poor, still racially inscribed, owed much to earlier aerospace ventures that depended on clearing them out. Populations deemed extraneous no longer required formal segregation to render them invisible. In areas requisitioned for space installations, the literal reshaping and remapping—terraforming—of the land obliterated their provenance.

The impact studies and strongly approving contemporaneous accounts of this reshaping of the South are peculiarly silent on the concurrent existence of a civil rights movement whose aims aerospace simultaneously recognized and evaded. A black Huntsville physician recalled "how the federal government was pumping money into the city and how the city was calling itself the Space Capital of the Universe. . . . But little of that money seemed to come our way."[77] Instead, investments in the technology of the industrial park, the mall, and the suburban tract made a *renewed* South a reality but left moot the actualization of a New South. Brokerage between elites and the federal government led to a resolution that rebuked the region's violent history. When this past nevertheless lived on in the behavior of unreconstructed rebels and, more important, in the unresolved issues forced into consciousness by the civil rights movement, it was perceived as disturbing a universalized, color-blind order that had already been negotiated.

The riposte by black writers and critics of speculative fiction has been to view racism as a technology embedded in the broader craft of the Space Age. History gives this view some heft: slavery, transatlantic abduction, and medical experimentation on blacks rivaled fiction. These practices applied specific technologies to achieve their ends. In Norman Mailer's words, "The real had become more fantastic than the imagined."[78]

Segregation survived the Space Age, emptied of racial content in America's renovated cities and suburbs. Aerospace had continued the contouring of southern land and the expulsion of populations begun decades earlier with military bases and New Deal infrastructural projects. Federal requirements had ended the exclusion of African Americans from such ventures, but inclusion had ceased to be their only goal. This essay breaks new ground as it links histories: southern, African American, and technology. It addresses the roots of the Sun Belt as the southern aerospace economy created a template for national development. Science fiction leaves its own spectral trace as disseminated by astronauts, theme park magnates, and funk musicians alike. There is little reason to suspect that the Space Age South is the last of the New South. The continual remaking of New Souths suggests that technology has not succeeded in bringing the South's restless patterns of reconstruction to closure. New Souths, of course, not only gesture to the future but show the future as always incomplete and in need of regeneration.

Notes

1. Gerard DeGroot, *Dark Side of the Moon: The Magnificent Madness of the American Lunar Quest* (New York: New York University Press, 2006), 66; Emory O. Jackson, "The Tip-Off," *Atlanta Daily World*, November 8, 1957, 4, 5; Benjamin E. Mays, "My View: If

Russia Is Ahead of Us," *Pittsburgh Courier*, December 21, 1957; Victor Calverton, "Blames Race Prejudice for U.S. Scientific Lag," *Chicago Defender*, November 9, 1957, 2.

2. On terraforming, see Carl Sagan, *Pale Blue Dot* (New York: Random House, 1994), 338–348.

3. The canonical work in the huge literature on the New South is C. Vann Woodward, *Origins of the New South* (Baton Rouge: Louisiana State University Press, 1951). Recent titles include John B. Boles and Bethany L. Johnson, eds., *Origins of the New South: Fifty Years Later: The Continuing Influence of a Historical Classic* (Baton Rouge: Louisiana State University Press, 2003); James C. Cobb, *Industrialization and Southern Society, 1877–1984* (Lexington: University Press of Kentucky, 2015); Reiko Hillyer, *Designing Dixie: Tourism, Memory, and Urban Space in the New South* (Charlottesville: University of Virginia Press, 2015); J. William Harris, *The New South: New Histories* (New York: Routledge, 2008); Robert H. Zieger, ed., *Life and Labor in the New New South* (Gainesville: University Press of Florida, 2012); Andrew Zimmerman, *Alabama in Africa: Booker T. Washington, the German Empire, and the Globalization of the New South* (Princeton, NJ: Princeton University Press, 2010).

4. Important accounts published by NASA include Steven J. Dick and Roger D. Launius, eds., *Societal Impact of Spaceflight* (Washington, DC: Government Printing Office [hereafter cited as GPO], 2007); Andrew J. Dunar and Stephen P. Waring, *Power to Explore: A History of Marshall Space Flight Center, 1960–1990* (Washington, DC: GPO, 1999); Charles D. Benson, *Moonport: A History of Apollo Launch Facilities and Operations* (Washington, DC: GPO, 1978); NASA Historical Staff, *Astronautics and Aeronautics* (Washington, DC: GPO, 1965).

5. These include Lillian Levy, ed., *Space: Its Impact on Man and Society* (New York: Norton, 1965); Annie Mary Hartsfield, *NASA Impact on Brevard County: Summary Report* (Tallahassee: Institute for Social Research, Florida State University, 1966); Loyd S. Swenson, "The Fertile Crescent: The South's Role in the Space Program," *Southwestern Historical Quarterly* 71 (1967–68): 377–379; Raymond Augustine Bauer, Richard S. Rosenbloom, Laure M. Sharp, and American Academy of Arts and Sciences Committee on Space Efforts and Society, *Second-Order Consequences: A Methodological Essay on the Impact of Technology* (Cambridge, MA: MIT Press, 1969).

6. Richard Paul and Steven Moss, *We Could Not Fail: The First African Americans in the Space Program* (Austin: University of Texas Press, 2015); Margot Lee Shetterly, *Hidden Figures: The American Dream and the Untold Story of the Black Women Who Helped Win the Space Race* (New York: William Morrow, 2016); De Witt Douglas Kilgore, *Astrofuturism: Science, Race and Visions of Utopia in Space* (Philadelphia: University of Pennsylvania Press, 2003); Neil M. Maher, *Apollo in the Age of Aquarius* (Cambridge, MA: Harvard University Press, 2017).

7. Quoted in James O. Farmer Jr., "A Collision of Cultures: Aiken, South Carolina, Meets the Nuclear Age," in *Proceedings of the South Carolina Historical Association* (1995), 41.

8. Morgan Fitz, "Hydrogen Bomb Explodes Boom Dreams of Town," *Ironwood Daily Globe*, January 13, 1951, 2; Deborah J. Holland, "Steward of World Peace, Keeper of Fair

Play: The American Hydrogen Bomb and Civil Rights, 1945–1954" (PhD diss., Northwestern University, 2002), 50–52; "Truman to Get Complaints of Bias at S.C. Atomic Project," *Baltimore Afro-American*, July 7, 1951, 1; Abby J. Kinchy, "African Americans in the Atomic Age: Postwar Perspectives on Race and the Bomb, 1945–1967," *Technology and Culture* 50 (April 2009): 309, 310; Vincent J. Intondi, *African Americans against the Bomb: Nuclear Weapons, Colonialism, and the Black Freedom Movement* (Stanford, CA: Stanford University Press, 2015).

9. Holland, "Steward of World Peace," 53, 43–160; Kari Frederickson, *Cold War Dixie Militarization and Modernization in the American South: Politics and Culture in the Twentieth-Century South* (Athens: University of Georgia Press, 2016), 49.

10. Quoted in Holland, "Steward of World Peace," 77; "Atom Plant Hiring Scored as Biased," *New York Times*, June 27, 1951, 24; "A.V.C. Scores Bias at Atomic Plants," *New York Times*, July 29, 1951, 47; "Hydrogen Bomb Unit Accused of Race Bias," *New York Times*, September 19, 1951; 20C; P. Trussell, "Job Bias Charged [at] Atomic Project," *New York Times*, April 18, 1952, 9.

11. Rufus Wells, "The Real A-Bomb Secret as Found at Oak Ridge," *Baltimore Afro-American*, June 23, 1951, 1, 2; Jack E. Wood to M. M. Wyatt, General Office Files, Papers of the NAACP, Library of Congress.

12. Holland, "Steward of World Peace," 216; "Ask Ban on Discrimination in Hydrogen Bomb Project," *Amsterdam News*, December 16, 1950, 28.

13. L. D. Reddick, "Africa, the Confederate Myth and the New Frontier," speech delivered at Coppin State Teachers College, Baltimore, April 22, 1961, 1, copy in Dr. Martin Luther King, Jr. Archive, Howard Gotlieb Archival Research Center, Boston University.

14. Allen J. Matusow, *The Unraveling of America: A History of Liberalism in the 1960s* (New York: Harper and Row, 1984), 9, 11, 12; Dale Carter, *The Final Frontier: The Rise and Fall of the American Rocket State* (London: Verso, 1988), 122–127; Adlai Stevenson to E. J. Ryan, October 23, 1957, in *The Papers of Adlai Stevenson*, vol. 7, ed. Walter Johnson (Boston: Little, Brown, 1977), 91.

15. Michael Curtin, *Redeeming the Wasteland: Television Documentary and Cold War Politics* (New Brunswick, NJ: Rutgers University Press, 1995), 29–30; Wayne J. Urban, *More Than Science and Sputnik: The National Defense Education Act of 1958* (Tuscaloosa: University of Alabama Press, 2010); Andrew Hartman, *Education and the Cold War: The Battle for the American School* (New York: Palgrave Macmillan, 2008); Gerald Horne, *Black and Red: W. E. B. Du Bois and the Afro-American Response to the Cold War* (Albany, NY: SUNY Press, 1986); Mary L. Dudziak, *Cold War Civil Rights: Race and the Image of American Democracy* (Princeton, NJ: Princeton University Press, 2000); Thomas Borstelmann, *The Cold War and the Color Line: American Race Relations in the Global Arena* (Cambridge, MA: Harvard University Press, 2001).

16. Gilbert C. Fite, *Richard B. Russell, Jr., Senator from Georgia* (Chapel Hill: University of North Carolina Press, 1991), 346, 363, 442, 443; United States Senate, 94th Congress, 2nd sess., *Committee on Aeronautics and Space Sciences, 1958–1976* (Washington, DC: GPO, December 30, 1976).

17. Statement by Y. K. Fyodorov at the Scientists Club press conference, Moscow, reported April 16, 1961, in Nikolai Tsymbal, ed., *The First Man in Space: The Record of*

Yuri Gagarin's Historic First Venture into Cosmic Space (New York: Cross Currents Press, 1961), 50; George Skorov, "Ivan Potekhin: Man, Scientist, and Friend of Africa," *Journal of Modern African Studies* 2, no. 3 (1964): 444–447.

18. Richard Paul, "How NASA Joined the Civil Rights Revolution, *Air & Space Magazine*, March 2014, http://www.airspacemag.com/history-of-flight/how-nasa-joined-civil-rights-revolution-180949497/?; Central Intelligence Agency, National Intelligence Estimate, November 5, 1961, 11, https://archive.org/stream/MissileGapCIAFiles/CIA%20DI%20ONE%20National%20Intelligence%20Estimate%2011-5-61%2C%20Soviet%20Technical%20Capabilities%20in%20Guided%20Missiles%20and%20Space%20Vehicles%2C%2025%20April%201961._djvu.txt.

19. "Astronaut Crosses 44 Africa Points," *Los Angeles Sentinel*, May 30, 1963, A2; "Cooper Relays 'Hello' to Africa from Space Craft," *Chicago Defender*, May 18, 1963, 3.

20. "JFK Tells Africa Leaders Confab Inspires Others," *Chicago Daily Defender*, May 23, 1963, 3; Richard D. Mahoney, *JFK: Ordeal in Africa* (New York: Oxford University Press, 1983), 30, 31; L. Gordon Cooper Jr. interviewed by Roy Neal, Pasadena, CA, May 21, 1998, NASA, Johnson Space Center Oral History Project, https://www.jsc.nasa.gov/history/oral_histories/CooperLG/CooperLG_5-21-98.htm; Buzz Aldrin and Wayne Warga, *Return to Earth*, 1st ed. (New York: Random House, 1973), 53–54, 75.

21. Erik Bergaust, *Rocket City, U.S.A.: From Huntsville, Alabama to the Moon* (New York: Macmillan, 1963), 25; Joseph A. Fry, *Dixie Looks Abroad: The South and U.S. Foreign Relations, 1789–1973* (Baton Rouge: Louisiana State University Press, 2002), 240; Walter A. McDougall, *The Heavens and the Earth: A Political History of the Space Age*, 2nd ed. (Baltimore: Johns Hopkins University Press, 1997), 376.

22. "Gulf Coast Aided by Moon Program," *New York Times*, October 6, 1963, 70; Jack Langguth, "Space Boom Stirs Louisiana Bayous," *New York Times*, September 8, 1963, 55; Carter, *Final Frontier*, 203; Fry, *Dixie Looks Abroad*, 238; Kevin M. Brady, "NASA Launches Houston into Orbit: The Economic and Social Impact of the Space Agency on Southeast Texas, 1961–1969," in *Societal Impact of Spaceflight*, ed. Steven J. Dick and Roger D. Launius (Washington, DC: NASA History Division, 2007), 458–459.

23. Paul Gaston, *The New South Creed: A Study in Southern Mythmaking* (Montgomery, AL: New South Books, 1970), 252.

24. Shaila Dewan, "When the Germans, and Rockets, Came to Town," *New York Times*, December 31, 2007, A11; David F. Noble, *The Religion of Technology: The Divinity of Man and the Spirit of Invention* (New York: Knopf, 1997), 124, 128; Sonnie Wellington Hereford III and Jack D. Ellis, *Beside the Troubled Waters: A Black Doctor Remembers Life, Medicine, and Civil Rights in an Alabama Town* (Tuscaloosa: University of Alabama Press, 2011), 88, 89.

25. Quoted in Noble, *Religion of Technology*, 126.

26. Kilgore, *Astrofuturism*, 168.

27. Roger D. Launius, "Interpreting the Moon Landings: Project Apollo and the Historians," *History and Technology* 22 (September 2006): 231; Gerard DeGroot, "The Dark Side of the Moon," *History Today* 57 (March 2007): 11–17; Remarks of Senator John F. Kennedy on Nikita Khrushchev's visit to the United States, National Kennedy for President Clubs press release, n.d., John F. Kennedy Pre-presidential Papers, 1946–1960,

Senate Files, Speeches and the Press, John F. Kennedy Presidential Library, Boston; Robert D. Dean, *Imperial Brotherhood: Gender and the Making of Cold War Foreign Policy* (Amherst: University of Massachusetts Press, 2001), 169, 170.

28. Lyndon B. Johnson, "The New World of Space," speech, Proceedings of the Second National Conference on the Peaceful Uses of Space (Washington, DC: GPO, 1962).

29. Alan Shepard and Deke Slayton, *Moon Shot: The Inside Story of America's Race to the Moon* (Atlanta: Turner Publishing, 1994), 165; Taylor Branch, *Parting the Waters: America in the King Years, 1954–63* (New York: Simon and Schuster, 1988), 863–864.

30. Robert A. Divine, "Lyndon B. Johnson and the Politics of Space," in *The Johnson Years: Vietnam, the Environment, and Science*, ed. Robert A. Divine (Lawrence: University Press of Kansas, 1987), 233; Brady, "NASA Launches Houston," 459.

31. Norman Mailer, *Of a Fire on the Moon* (Boston: Little, Brown, 1969), 9; Matthew D. Lassiter and Kevin M. Kruse, "The Bulldozer Revolution: Suburbs and Southern History since World War II," *Journal of Southern History* 75 (August 2009): 691–796; William S. Ellis, "Space Crescent III: The Wide Blue Porkbarrel," *Nation*, October 26, 1964, 276.

32. Lori C. Walters, "Beyond the Cape: An Examination of Cape Canaveral's Influence on the City of Cocoa Beach, 1950–1963," *Florida Historical Quarterly* 87 (Fall 2008): 235–257; Re Mr. Henri Langwirth, *Congressional Record* 149, pt. 4, February 25, 2003, to March 10, 2003, 5566; Jerrell H. Shofner, *History of Brevard County* (Brevard County, FL: Brevard County Historical Commission, 1995), 151; James Loewen, *Sundown Towns: A Hidden Dimension of American Racism* (New York: New Press, 2005); "Race Fight Looms Over Della Reese's Fla. Pact," *Chicago Daily Defender*, October 8, 1962, 16; *Jazz News*, October 31, 1962, 6.

33. NAACP press release, "Arrests of Youth Spur Desegregation Drive," June 22, 1962, NAACP Papers, Part 20: White Resistance and Reprisals, 1956–1965; "Protest to JFK About Bias at Cape Canaveral," *Chicago Daily Defender*, June 12, 1962, 3; "Ella Missed Florida But Not Della Who Integrated Clubs in Missile Land," *Chicago Daily Defender*, October 29, 1962, 17.

34. Izzy Rowe, "Izzy Rowe's Notebook," *Pittsburgh Courier*, November 3, 1962, 14.

35. Advertisement for Canaveral Acres, *Ebony*, May 1963, 10; Paul and Moss, *We Could Not Fail*, 44; Wade Arnold, ed., *Florida's Space Coast* (Charleston, SC: Arcadia Publishing, 2009); Roy Wilkins to John F. Kennedy, November 20, 1962, NAACP Papers, Part 05: Campaign against Residential Segregation, Supplement: Residential Segregation, General Office Files, 1956–1965.

36. McDougall, *Heavens and the Earth*, 376.

37. Lassiter and Kruse, "Bulldozer Revolution," 695; N. D. B. Connolly, *A World More Concrete: Real Estate and the Remaking of Jim Crow South Florida* (Chicago: University of Chicago Press, 2014), 150–151.

38. "NAACP Charges Bias at Cape Canaveral," *Atlanta Daily World*, March 24, 1963, A1, A4; Peter Dodd, *Social Change in Space-Impacted Communities* (Cambridge, MA: Committee on Space of the American Academy of Arts and Sciences, 1964), 20–21; Sallie Middleton, "Space Rush: Local Impact of Federal Aerospace Programs on Brevard and Surrounding Counties," *Florida Historical Quarterly* 87 (Fall 2008): 275; Elaine M. Stone,

Brevard County: From the Cape of the Canes to Space Coast (Northridge, CA: Windsor Publications, 1988), 67.

39. Nash quoted in Clayborne Carson, *In Struggle: SNCC and the Black Awakening of the 1960s* (Cambridge, MA: Harvard University Press, 1981), 13; Michael E. Latham, *Modernization as Ideology: American Social Science and "Nation Building" in the Kennedy Era* (Chapel Hill: University of North Carolina Press, 2000); Christopher T. Fisher, "'The Hopes of Man': The Cold War, Modernization Theory, and the Issue of Race in the 1960s" (PhD diss., Rutgers University, 2002); H. L. T. Quan, *Growth against Democracy: Savage Developmentalism in the Modern World* (Lanham, MD: Lexington Books, 2012).

40. "The Spell of Southern Warmth: *Life*'s Gulf Coast and Florida Tour," *Life*, February 2, 1962, 50–72.

41. Robert G. Scharff, *Louisiana's Loss, Mississippi's Gain: A History of Hancock County, Mississippi, from the Stone Age to the Space Age* (Lawrenceville, VA: Brunswick, 1999), 588; Bergaust, *Rocket City*, 168; "La. Rejects Federal Money Because It Can't Segregate," *New Pittsburgh Courier*, April 18, 1964, 18.

42. Wallace quoted in Michael J. Neufeld, *Von Braun: Dreamer of Space, Engineer of War* (New York: Alfred A. Knopf in association with the National Air and Space Museum, Smithsonian Institution, 2007), 396; Lady Bird Johnson quoted in Ernest R. May and Guian McKee, eds., *Lyndon Johnson: Toward the Great Society, February 1, 1964–May 31, 1964* (New York: W. W. Norton, 2007), 89.

43. William S. Ellis, "Space Crescent II: The Brain Ghettos," *Nation*, October 19, 1964, 239–241; Neufeld, *Von Braun*, 386; Ernst Stuhlinger and Frederick I. Ordway III, *Wernher Von Braun, Crusader for Space: A Biographical Memoir* (Malabar, FL: Krieger, 1994), 187.

44. Bergaust, *Rocket City*, 59, 212; Monique Laney, *German Rocketeers in the Heart of Dixie: Making Sense of the Nazi Past during the Civil Rights Era* (New Haven, CT: Yale University Press, 2015); Elise H. Stephens, *Historic Huntsville: A City of New Beginnings* (Woodland Hills, CA: Windsor, 1984), 124; Ben A. Franklin, "Von Braun Fights Alabama Racism: Scientist Warns State U.S. Might Close Space Center," *New York Times*, June 14, 1965, 39; Kim McQuaid, "Racism, Sexism, and Space Ventures: Civil Rights at NASA in the Nixon Era and Beyond," in Dick and Launius, *Societal Impact of Spaceflight*, 266.

45. Bergaust, *Rocket City*, 59.

46. McDougall, *Heavens and the Earth*, 399; Fred Powledge, "Demonstrations Atop Unisphere and Giant Orange Considered," *New York Times*, April 21, 1964, 30.

47. Diggs's letters to the secretaries of state and defense, Records of the House Foreign Affairs Subcommittee on Africa, Box 191, Folder 22, Charles Diggs Papers, Moorland-Spingarn Collection, Howard University; Memorandum from Charles E. Johnson and Ulric Haynes of the NSC Staff to President Johnson, July 13, 1965, US Department of State, *Foreign Relations of the United States*, vol. 26, 1964–1968, *Africa*, 1031; Charles F. Baird to Ivanhoe Donaldson, February 14, 1967, Papers of the Student Nonviolent Coordinating Committee, Series VIII, General Files, Reel 51, University Microfilms International, 2014.

48. Herbert Roof Northrup, *The Negro in the Aerospace Industry* (Philadelphia: Wharton School of Finance and Commerce, University of Pennsylvania, 1968), 38.

49. US Senate, 93rd Cong., 1st sess., Senate Committee on the Judiciary, Subcommittee on Constitutional Rights, *Military Surveillance of Civilians*, part 1 (Washington, DC: GPO, 1973), 30; "Aerojet History," http://www.aerojet.com/about/history.php; "1984 Supercar for 1968 Superfuzz," *Los Angeles Free Press*, December 15–22, 1967, 1.

50. Buzz Aldrin, *Magnificent Desolation: The Long Journey Home from the Moon* (New York: Harmony Books, 2009), 9; Mailer, *Of a Fire on the Moon*, 339.

51. Edward Wynne to Meyer Feldman, Deputy Special Counsel to the President, October 10, 1962, John F. Kennedy Library, quoted in Lynn Spigel, *Welcome to the Dreamhouse: Popular Media and Postwar Suburbs* (Durham, NC: Duke University Press, 2001), 176n21; Brenda Gayle Plummer, *In Search of Power: African Americans in the Era of Decolonization, 1956–1974* (New York: Cambridge University Press, 2013), 242; Cathleen S. Lewis, "Arnaldo Tamayo Mendez and Guion Bluford: The Last Cold War Race Battle," NASA in the Long Civil Rights Movement Conference, Huntsville, AL, March 16, 2017.

52. Shofner, *History of Brevard County*, 121–123; Carter, *Final Frontier*, 227.

53. Bernard Weinraub, "Bustle at Cape Bypasses the Hungry," *New York Times*, July 14, 1969, 1, 23; Steve Huntley, "SCLC in Space Parley," *Chicago Daily Defender*, July 16, 1969, 4; "Giant Leap for Mankind?" *Ebony*, September 1969, 58.

54. DeGroot, *Dark Side of the Moon*, 234–235.

55. *The Black Panther*, January 25, 1969, page numbers illegible.

56. John Rieder, *Early Classics of Science Fiction: Colonialism and the Emergence of Science Fiction* (Middletown, CT: Wesleyan, 2012), 18, 113.

57. Roger D. Launius, "NASA and the Decision to Build the Space Shuttle, 1969–72," *Historian* 57, no. 1 (1994): 26.

58. Gil Scott-Heron, *Now and Then: The Poems of Gil Scott-Heron* (Edinburgh: Payback Press, 2000), 21; Thomas A. Johnson, "Blacks and Apollo: Most Couldn't Have Cared Less," *New York Times*, July 27, 1969, E6.

59. *News-Tribune* (Rome, GA), February 2, 1971, 12.

60. Roger D. Launius, "Heroes in a Vacuum: The Apollo Astronaut as Cultural Icon," *Florida Historical Quarterly* 87 (Fall 2008): 202–204; Frank Borman, *Countdown: An Autobiography* (New York: Silver Arrow Books, 1988), 207.

61. Kriston Capps, "Of Course There Are Black Stormtroopers in Star Wars," *Atlantic*, November 28, 2014, http://www.theatlantic.com/entertainment/archive/2014/11/of-course-there-are-black-stormtroopers-in-star-wars/383259/; Alondra Nelson, "Introduction: Future Texts," *Social Text* 20 (Summer 2002): 6.

62. Ramzi Fawaz, "Space, That Bottomless Pit: Planetary Exile and Metaphors of Belonging in American Afrofuturist Cinema," *Callaloo* 35, no. 4 (2012): 1199.

63. Isiah Lavender III, *Race in American Science Fiction* (Bloomington: Indiana University Press, 2011), 20; Constance Penley, *NASA/Trek: Popular Science and Sex in America* (New York: Verso, 1997), 88.

64. Paul, "How NASA Joined the Civil Rights Revolution"; Matthew D. Tribbe, *No Requiem for the Space Age: The Apollo Moon Landings and American Culture* (New York: Oxford University Press, 2014), 38; William S. Ellis, "The Space Crescent: Moon Boom," *Nation*, October 12, 1964, 211; John W. Finney, "NASA Is Training Negroes for Jobs: But Qualified Applicants Are Difficult to Find," *New York Times*, May 31, 1964, 54.

65. McQuaid, "Racism, Sexism," 157, 169; Andrea Brunais, "Memories of a Child of the Space Program," *FHC Forum* 20 (Winter 1997–98): 15; Shetterly, *Hidden Figures*; Nathalia Holt, *Rise of the Rocket Girls: The Women Who Propelled Us, from Missiles to the Moon to Mars* (New York: Little, Brown, 2016); Katherine Johnson Receives Presidential Medal of Freedom, https://www.nasa.gov/image-feature/langley/katherine-johnson-receives-presidential-medal-of-freedom; *Hidden Figures*, dir. Theodore Melfi, Fox 2000 Pictures, 2016.

66. Holt, *Rise of the Rocket Girls*, 166; "Space Agency Fires Black Woman Who Attacked Bias," *Jet*, November 15, 1973, 6; Ruth Harris, *Harlem Princess: The Story of Harry Delaney's Daughter* (New York: Vantage Press, 1991), 270; Stephanie Nolen, *Promised the Moon: The Untold Story of the First Women in the Space Race* (New York: Thunder's Mouth Press, 2002), 276.

67. Walters, "Beyond the Cape," 252; Carter, *Final Frontier*, 192; Kilgore, *Astrofuturism*, 52–54, 57–60; M. G. Lord, *Astro Turf: The Private Life of Rocket Science* (New York: Walker, 2005), 111–115; John W. Finney, "Astronauts Cited on Motel's Action: Negro Was Barred in Florida Inn, Young Charges," *New York Times*, March 5, 1964, 27.

68. Cedric Robinson, *Forgeries of Memory and Meaning: Blacks and the Regimes of Race in American Theater and Film before World War II* (Chapel Hill: University of North Carolina Press, 2012), 54.

69. Neufeld, *Von Braun*, 285–287; Kilgore, *Astrofuturism*, 52–54, 57–60; Middleton, "Space Rush," 272.

70. Neufeld, *Von Braun*, 452–462.

71. Andre M. Carrington, *Speculative Blackness: The Future of Race in Science Fiction* (Minneapolis: University of Minnesota Press, 2016), 2; Lisa Yaszek, "Afrofuturism, Science Fiction, and the History of the Future," *Socialism and Democracy* 20, no. 3 (2006): 47; Kodwo Eshun, "Further Considerations of Afrofuturism," *CR: The New Centennial Review* 3 (Summer 2003): 299.

72. Sandra Jackson and Julie E. Moody-Freeman, eds., *The Black Imagination, Science Fiction, Futurism and the Speculative* (New York: Peter Lang, 2011); Adilifu Nama, *Black Space: Imagining Race in Science Fiction Film* (Austin: University of Texas Press, 2008); Isiah Lavender, ed., *Black and Brown Planets: The Politics of Race in Science Fiction* (Jackson: University Press of Mississippi, 2014); Sandra M. Grayson, *Visions of the Third Millennium: Black Science Fiction Novelists Write the Future* (Trenton, NJ: Africa World Press, 2003); Andre M. Carrington, *Speculative Blackness*; Eshun, "Further Considerations of Afrofuturism," 298; Timothy A. Spaulding, *Re-forming the Past: History, the Fantastic, and the Postmodern Slave Narrative* (Columbus: Ohio State University Press, 2005); Patricia Melzer, *Alien Constructions: Science Fiction and Feminist Thought* (Austin: University of Texas Press, 2006); Jeffrey A. Tucker, *A Sense of Wonder: Samuel R. Delany, Race, Identity, and Difference* (Middletown, CT: Wesleyan University Press, 2004); Darieck Scott, *Extravagant Abjection: Blackness, Power, and Sexuality in the African American Literary Imagination* (New York: New York University Press, 2010).

73. Nama, *Black Space*, 153.

74. Ken McLeod, "Space Oddities: Aliens, Futurism and Meaning in Popular Music," *Popular Music* 22, no. 3 (2003): 337–355; Chris Richards, "Smithsonian Acquires

Parliament-Funkadelic Mothership," *Washington Post*, May 18, 2011, https://www.washingtonpost.com/lifestyle/style/smithsonian-acquires-parliament-funkadelic-mothership/2011/05/18/AFHMvj6G_story.html; Roger D. Launius, "Abandoned in Place: Interpreting the U.S. Material Culture of the Moon Race," *Public Historian* 31 (August 2009): 9–38; Ytasha Womack, *The World of Black Sci-Fi and Fantasy Culture* (Chicago: Lawrence Hill Books, 2013), 33–35.

75. Eshun, "Further Considerations of Afrofuturism," 294; Daniel Kreiss, "Appropriating the Master's Tools: Sun Ra, the Black Panthers, and Black Consciousness," *Black Music Research Journal* 28 (Spring 2008): 61; Fawaz, "Space," 1104–1119.

76. Elizabeth Woyke, "Billionaires in Space," *Forbes*, March 5, 2008, www.forbes.com/2008/03/05/space-tourists-billionaires-tech-billionaires08-cx_ew_0305space.html.

77. Sonnie Wellington Hereford III and Jack D. Ellis, *Beside the Troubled Waters: A Black Doctor Remembers Life, Medicine, and Civil Rights in an Alabama Town* (Tuscaloosa: University of Alabama Press, 2011), 88.

78. Nabeel Zuberi, "Is This the Future? Black Music and Technology Discourse," *Science Fiction Studies* 34, no. 2 (2007): 283–300; Womack, *Afrofuturism*; LaRose Davis, "Future Souths, Speculative Souths, and Southern Potentialities," *PMLA* 131, no. 1 (2016): 191–192; Mailer, *Of a Fire on the Moon*, 141.

Figure 1. Langley Research Center's Equal Employment Opportunity Committee meeting to review the center's Affirmative Action Program in 1965. *Seated from left*: Katherine G. Johnson, Lawrence W. Brown, and J. Norwood Evans; *Standing from left*: John J. Cox and Edward T. Maher. Courtesy of NASA.

Figure 2. Marshall Space Flight Center mathematician Billie Robertson running a real-time simulation of Translunar Injection (TLI) Go-No-Go for Apollo 17 on November 27, 1972. Robertson began her career working for the Wernher von Braun team in 1952. Courtesy of NASA.

Figure 3. Marshall Space Flight Center "Negro Recruiter" Charles T. Smoot. In the fall of 1963, Smoot helped establish a cooperative program with Southern University in Baton Rouge, Louisiana. Courtesy of NASA.

Figure 4. Jeanette Scissum at her desk in the Marshall Space Sciences Laboratory in 1967. In 1975, Scissum authored an essay for the National Technical Association titled "Equal Employment Opportunity and the Supervisor: A Counselor's View," which argued that many discrimination complaints could be avoided "through adequate and meaningful communication." Courtesy of NASA.

Top left: Figure 5. STS-8 mission specialist Guion S. Bluford Jr. during a July 13, 1983, pr
conference. Bluford became the United States' first African American astronaut when he w
selected in August 1979. Courtesy of NASA.

Top right: Figure 6. Katherine G. Johnson at the Langley Research Center on September 8, 19
In 2015, President Barack Obama awarded Johnson the Presidential Medal of Freedom. Co
tesy of NASA.

Above: Figure 7. Director of the Marshall Space Flight Center Wernher von Braun and Alaba
A&M University president Richard D. Morrison sign a cooperative agreement in Novem
1968. Courtesy of NASA.

Figure 8. Langley Research Center engineer Thomas A. Byrdsong checks the Apollo/Saturn IB ground-wind-loads model in the Langley Transonic Dynamics Tunnel on March 2, 1963. Courtesy of NASA.

Figure 9. Governor of Alabama George Wallace (*left*), NASA administrator James Webb (*center*), and Marshall Space Flight Center director Wernher von Braun (*right*) during a tour of the Marshall Space Flight Center on June 8, 1965. Courtesy of NASA.

Figure 10. Clyde Foster was a charter member of the Army Ballistic Missile Agency team transferred to NASA in July 1960. Over his career, Foster worked in the Marshall Space Flight Center Computation Laboratory; was mayor of Triana, Alabama; developed a computer science program at Alabama A&M University; and was named director of Marshall's Equal Employment Opportunity Office in 1975. Courtesy of NASA.

Figure 11. Visit by the head of the National Urban League, Whitney Young Jr., to Marshall in May 1966. From left: W. H. Hollins (Alabama A&M University), Milton Cummings (president of Brown Engineering), Whitney Young Jr., Dave Newby (Marshall Space Flight Center), and Richard D. Morrison (president of Alabama A&M University). Courtesy of NASA.

Figure 12. On November 16, 1964, Hobart Taylor Jr., the executive vice chairman of the President's Committee on Equal Employment Opportunity (PCEEO), visited the Marshall Space Flight Center to review its progress on equal employment opportunity. *From left*: Hobart Taylor Jr., Alfred Hodgson (NASA's Equal Employment chief compliance officer), and Karl Heimburg (director of NASA Marshall Test Laboratory). Courtesy of NASA.

Figure 13. Groundbreaking of science building at Miles College (Birmingham, Alabama) on November 22, 1964. Marshall Space Flight Center director Wernher von Braun spoke at the event at the request of Miles's president, Lucius H. Pitts. Courtesy of NASA.

Figure 14. NASA Equal Employment Opportunity Office director Ruth Bates Harris during an October 13, 1972, visit to the Marshall Space Flight Center. *From left*: Robert E. Shurney, Ruth Bates Harris, and Inellia Sullivan. Courtesy of NASA.

Above: Figure 15. Students working in the satellite tracking station at Alabama A&M University in 1964. Courtesy of NASA.

Left: Figure 16. Edward (Ed) J. Dwight Jr. was selected by President John F. Kennedy in 1961 to be the first African American astronaut. Dwight resigned from the program and the air force in 1966. Courtesy of NASA.

Figure 17. Mercury 13 astronaut Jerrie Cobb posing next to the Mercury capsule in the early 1960s. Cobb was a participant in Randy Lovelace's First Lady Astronaut Trainees (FLAT) program. Courtesy of NASA.

Figure 18. Mary Helen Johnston, Carolyn Griner, and Ann Whitaker photographed in the Marshall Space Flight Center's Neutral Buoyancy Simulator in 1976. The three women were conducting training related to the Concept Verification Testing of Materials for Science Payloads. *From left*: Johnston, Griner, and Whitaker. Courtesy of NASA.

Figure 19. *Star Trek* star Nichelle Nichols meeting with female scientists at the Marshall Space Flight Center. *From left around the table*: Nichelle Nichols, Mary Helen Johnston, Ann Whitaker, and Doris Chandler. Courtesy of NASA.

Figure 20. In 1978, NASA selected the first six female astronauts to be mission specialists to fly on the space shuttle. On June 18, 1978, Sally Ride became the first American woman to fly in space when she flew on STS-7. *From left*: Rhea Seddon, Anna L. Fisher, Judith A. Resnik, Shannon W. Lucid, Sally K. Ride, and Kathryn D. Sullivan. Courtesy of NASA.

Figure 21. Illustration of group in chains landing on Moon. *The Black Panther: Black Community News Service*, July 26, 1969, p. 2 © 2019 Emory Douglas / Artists Rights Soci (ARS), New York. Photo courtesy of Art Resource, NY.

Figure 22. Illustration of pig hammering "White Only" sign on Moon. *The Black Panther: Black Community News Service*, August 23, 1969, p. 28. © 2019 Emory Douglas / Artists Rights Society (ARS), New York. Photo courtesy of Art Resource, NY.

Accommodating the Forces of Change

Civil Rights and Economic Development in Space Age Huntsville, Alabama

MATTHEW L. DOWNS

On May 15, 1965, James E. Webb, the director of the National Aeronautics and Space Administration (NASA), spoke to a group of University of Alabama alumni at the National Press Club in Washington, DC. Webb highlighted the economic impact of the space program on Alabama and Huntsville, specifically citing investments of $1.8 billion. Moreover, Webb noted, those funds meant associated economic growth across the region, including retail sales, tax revenue, and even increased tourism. Webb touted the centrality of federal funds to the region's postwar prosperity and promised that such growth would continue as long as civic and business leaders demonstrated the "vision and the resourcefulness" to support that program.[1]

Webb's combination of praise and admonition spoke to long-seated concerns in Huntsville, a city that had become almost entirely dependent on federal investment for its economic future. Business leaders were well aware of the "great economic impact" of the space and missile programs, and, in fact, since World War II city leaders worked tirelessly to attract and maintain government defense projects, even as they tried to diversify the regional economy. But his words also hinted at a more immediate concern shared by their colleagues across the South. In May 1965, amid an ongoing civil rights struggle for racial equality, Webb's call for "vision and resourcefulness" suggested that the region's white civic and business leaders would have to adapt if they were to continue to benefit from federal largesse. In the words of the director of Huntsville's George C. Marshall Space Flight Center, the famed German rocket scientist Wernher von Braun, those leaders would have to "accommodate the forces of change" if Huntsville and the rest of the region hoped to preserve the prosperity brought by federal appropriations.[2]

This phrase, "accommodate the forces of change," used by von Braun in a 1965 speech to the Alabama State Chamber of Commerce, encapsulates the way in which many white civic and business leaders, and particularly those in Huntsville, approached the demands of the civil rights movement. In response to pressure by civil rights activists, white leaders offered compromise in the hopes of quieting protests, diverting national attention, placating federal representatives, and, above all, maintaining economic investments. As in Huntsville, such compromises helped parts of the modernizing South avoid the massive public confrontations that occurred in places like Birmingham and Selma, Alabama, while preserving enough of the status quo to forestall a larger transformation of the region's racial and social structures.

The story of the political and economic response by Huntsville's civic leadership to the local civil rights movement sits at the intersection of two historiographical trends in the history of the modern struggle for racial equality. Beginning in the 1980s, historians reached beyond the mainstream narrative of the movement, which focused on Martin Luther King Jr. and his efforts in Montgomery, Birmingham, and Selma, to understand the thousands of local struggles in which black men and women fought for equality. Robert J. Norrell's *Reaping the Whirlwind*, which focuses on Tuskegee, Alabama; Adam Fairclough's broad survey of the movement in Louisiana, *Race and Democracy*; Stephen Tuck's *Beyond Atlanta*; and other influential works have pushed scholars to think about the disparate protests and acts of courage that propelled the larger effort forward.[3] Huntsville's experience mirrors many of these community studies, with prominent local leaders and an engaged community that pressed for changes specific to the character of Jim Crow in the Rocket City. Locals were aware of and engaged in larger national conversations, but they were equally attuned to the problems and possibilities presented by Huntsville's unique economic and political mindset and the particular way that segregation had developed in the city.

Too, the story of Huntsville adds to the more specific history of the response by Sun Belt–style business interests to the intensifying pressure of the civil rights movement. Historians have long noted the participation of white southern business leaders in the response to civil rights protest. For instance, both Mills Thornton (in *Dividing Lines*) and Glenn Eskew (in *But for Birmingham*) recount the way that Birmingham's "moderate" white leadership stressed accommodation in the face of civil rights protests and the violent segregationist response.[4] But such works tend to focus most of their attention on the activists themselves.[5] The best and most comprehensive account of the response by Sun Belt business leaders to the movement is Elizabeth Jacoway and David R. Colburn's edited collection, *Southern Businessmen and Desegregation*, which includes a number of diverse studies, covering cities from Birmingham to Tampa, Florida, to Little Rock, Arkansas. While the authors' conclusions vary, the general theme of the work, as detailed by Jacoway in her introduction, is that business leaders played a crucial role in the way that the white community responded to the movement, whether to encourage economic growth, maintain a beneficial public image, or preserve the status quo. More recent works by Benjamin Houston on Nashville and Tracy K'Meyer on Louisville expand on and complicate this conclusion, noting the interaction between more "moderate" business interests and the continued demands of black activists and their allies. This essay, tracing similar interactions in Huntsville, speaks to this complicated history of local movements in the modernizing South.[6]

In many ways, the city of Huntsville is a perfect example of the Sun Belt, with rapid growth propelled by federal investment and civic boosterism. Hit hard by the Great Depression, which drove down agricultural prices and contributed to the decline of the textile industry, Huntsville's fortunes shifted when, in 1941, on the eve of World War II, the army located a chemical weapons arsenal and storage depot on farmland southwest of the city (a facility eventually referred to as Redstone Arsenal). Postwar demobilization threatened to return the city to financial straits, but the work of civic and business leaders and their connections in Washington, DC, helped to bring a revitalization at the hands of German rocket scientists who relocated to the then-defunct facility at Redstone to develop missile technology for the army. The site once again became the engine of Huntsville's economy; by the mid-1950s the arsenal and its web of contractors employed 12,000 with a payroll approaching $40 million. Civic leaders could say without exaggeration that "as Redstone grows, so grows Huntsville." The creation of NASA and the decision to locate the George C. Marshall Space Flight Center on the arsenal reservation to oversee a civilian rocket program added to the city's economic base. Once more, the city would benefit from an influx of federal funds, this time to NASA scientists, technicians, and their web of associated contractors.[7]

Such an overwhelming reliance on one "industry" understandably concerned the city's business leaders. As early as 1943, the business community pressed for economic diversification, arguing that the eventual return of a peacetime economy (then still a couple years away) would mean the end of the city's economic boom. The World War II–era arsenal's demobilization underlined those fears, and by the late 1940s a number of Huntsville and North Alabama civic organizations, including the regional North Alabama Associates and the Huntsville Industrial Expansion Committee, actively sought to bring in new industry. Boosters advertised the city, traveled to visit potential clients, and worked closely with the state and federal government in order to provide incentives and to clear obstacles to localized growth. Such efforts worked to attract investment, yet most of the industry attracted to the city relied, at least in part, on government contracts—the economy grew but failed to diversify. In a 1958 address to the Huntsville Rotary Club, Redstone's General Bruce Medaris echoed this ongoing concern, telling leaders to "continue your efforts to broaden the industrial base."[8]

By the 1960s, Huntsville was dependent on continued federal investment and was determined to expand its economic base; in both cases, the city needed to sell itself, and business and municipal leadership embraced the image and rhetoric of political and social moderation, even as their counterparts in other southern cities turned reactionary. Huntsville's boosters emphasized the city's racial harmony. In its literature, the industrial expansion committee advertised housing and school options for black residents and stressed that "Huntsville is a friendly and hospitable city, free of factionalism, of racial, religious, and political nature." In public statements, Mayor R. B. Searcy promised that "there are no problems in Huntsville . . . the Negroes and the white people had always gotten along well together." City leaders embraced what the historian William Chafe called the "progressive mystique," an "elusive" set of beliefs and assumptions that implied consensus, cooperation, and, in Chafe's words, "a pervasive commitment to civility" that masked the white community's paternalistic treatment of their black neighbors.[9]

In reality, segregation, prejudice, and their socioeconomic repercussions marked life for the black community. As Sonnie Hereford III remembered, he and other black Huntsvillians lived in a world "run by whites." They were refused entrance to downtown hotels, the public library, and white neighborhoods (unless employed as domestic workers), and they received inferior treatment in stores and restaurants. The only city school for black students, Councill, was crowded and without basic educational equipment or amenities. John Cashin remembered his own first experience of racism. Waiting at the window of a hobby shop connected to the city's YMCA in the hopes of buying model airplane parts, Cashin was doused by dirty mop water, a message from the manager that he was unwelcome; the incident showed him the "ugly meanness that could animate [segregation]." Both Cashin and Hereford learned that the best way to navigate segregated society was to avoid any situation that might lead to a confrontation with white Huntsvillians; as Chafe noted, civility for blacks assumed deference to whites, even if such deference was dishonest to the true feelings within the black community.[10]

The economic boom that Huntsville experienced during and after World War II suggested the possibility of improvement for African Americans, at least economically. Not only did federal investment flow into the city, but alongside such growth came an influx of new people, including German rocket scientists and a number of transplants from the North and West where racial prejudice was not as overt; Hereford remembered feeling that things might "get better." But Huntsville's boom did not spread evenly. The newcomers tended to, in the words of Hereford, "[go] along with the program," and black Huntsvillians were largely shut out of the economic opportunities that came with the development. Shelby Johnson wrote to Senator John Sparkman, complaining that skilled, educated black men and women were relegated to menial work; his own daughter, a Howard University–educated social worker, was rejected because "a white person would fill the job better." Sparkman told Johnson to be patient and wait for improvements. Those who did find work on the reservation remained in the most tenuous of positions, and when the arsenal began preparing to demobilize with the end of the war, black women were the first to be fired. Writing in 1960, S. W. Ellis captured the frustration many black Alabamians felt as economic opportunity passed them by: "Your so-called increased educational, business, and social opportunities of all our people does [sic] not [actually] stand up. What good is an education if the only work you can get is low-class labor? . . . Can one get a job at Hayes Aircraft on government contracts except as a janitor or aircraft washer? Can one get a job at Redstone Arsenal?"[11]

Sparkman's call for patience, and the language of the city's boosters promising a community prepared for and conducive to growth, implied that with prosperity would come improvement to lift all Huntsvillians. But in reality black life remained mired in Jim Crow. Hereford, who practiced medicine in the city in the 1950s, routinely visited the crowded shacks of the city's black neighborhoods, marked by leaking roofs, outhouses, and no ability to pay for basic medical care. Black workers remained a miniscule percentage of the city's government-related workforce, thanks in part to the fact that, as one black employee of the arsenal remembered, "We [blacks] didn't have a population with the prerequisites that would be needed to do this type of work." Those who did seek educational programs that would prepare individuals for work in the rocket labs faced the strictures of Jim Crow; Alabama A&M's leadership cautiously toed the color line and proved reluctant to offer an engineering program that might compete with the courses, reserved for whites, at the University of Alabama's Huntsville (UAH) branch.[12]

Huntsville's civic and business leaders renewed their efforts to diversify the city's economy just as black southerners escalated efforts to challenge racial inequality. The Montgomery bus boycott, which ended in 1957, brought national attention to Martin Luther King Jr., and to his organization, the Southern Christian Leadership Conference (SCLC). In 1960, black students from North Carolina A&M led sit-in protests in Greensboro, an act that sparked similar protests across the South.[13] The strategies developed by King and the boycott's leadership—nonviolent public protests calling attention to entrenched inequality—proved effective in putting pressure on civic leaders to respond. Efforts to maintain segregation in the face of such protests made Alabama's leaders national symbols of racist oppression.

Southern businessmen realized that a violent response by segregationist forces to peaceful protestors demanding freedom and equality created an image deeply unsettling to national corporations and potential investors. In many cases, moderate white business leaders concerned about such investment took the lead in shaping compromises designed to end public protests. Most famously, during the Birmingham movement of spring 1963, a group of industrialists and business executives hammered out a compromise that provided for desegregation of the downtown shopping district and a promise of future improvements in exchange for an end to the protests. As Sidney Smyer, a realtor who took the lead in seeking compromise with Martin Luther King Jr., noted, "If we're going to have a good business in Birmingham, we better change our way of living." The city's experience provided two important lessons for other civic leaders: hostile public confrontation attracted negative media portrayals that could damage a city's reputation, and efforts by moderate white leaders to find compromise might forestall further tensions.[14]

In other southern cities, white moderates sought to head off protests by addressing the most egregious aspects of Jim Crow before protests could start. In his study of Augusta, Georgia's, efforts, James Cobb noted that business leaders consciously worked to avoid becoming "vulnerable" to civil rights protests and worried publicly about the city's reputation among potential industrial clients; they eventually decided that "further resistance [to desegregation] would be futile and economically unwise." In Dallas, similar concerns over the city's image led black and white community leaders to create a biracial committee to stress "peaceful change." After "limited protests" over segregated restaurants and an Easter boycott of retail establishments, Dallas's biracial committee reached a compromise to desegregate lunch counters and hotels.[15]

When African Americans in Huntsville challenged the city's "progressive mystique" and actively pressed for change, the city's white leadership had to respond, and had to do so in a way that avoided a public confrontation that might tarnish the city's image. At the same time, those leaders were unwilling to embrace the full measure of equality that the black community demanded, given the climate of the city and the demands of their own constituents. Thus, Huntsville's response to the civil rights movement exemplifies the moderation of other Sun Belt cities, working to preserve opportunities for economic development and conceding to the demands of the movement when it became necessary to do so and in ways that failed to address deep-seated structural economic and social inequality.

Huntsville had a relatively brief history of civil rights activism. In the 1930s, students at Oakwood College led a strike to protest their all-white administration and forced campus leaders to hire more black administrators and faculty. John Cashin, a dentist who would become a central figure in the city's civil rights movement, led successful efforts to integrate both the municipal golf course and the library in the 1950s. But the city's movement began in earnest in January 1962 when, encouraged by the arrival of a young Congress of Racial Equality (CORE) staffer and veteran Freedom Rider named Hank Thomas, a group of African American students held sit-ins in the downtown business district. Two students were arrested, and the rest were denied service at Walgreens, Woolworth's, and a number of other lunch counters. Soon after, Cashin, Sonnie Hereford III, and other prominent black citizens formed the Community Service Committee (CSC) to oversee the protests and organize legal and community protection for the activists. From the start, the group's leaders understood the importance of reputation and federal money to Huntsville's well-being; as Hereford remembered, government contracts were a kind of "leverage" that ensured the city would "[try] to make things go as smoothly as possible" once desegregation seemed inevitable. As he wrote in a letter he sent to potential supporters, "[City officials] know that if the news reaches the outside and especially Washington, there will be outside pressure."[16]

The CSC escalated efforts to draw national attention to Huntsville's continued recalcitrance. When Martha Hereford and a visibly pregnant Joan Cashin were arrested (along with the Cashins' young daughter) during a sit-in at Walgreens, photos of the event appeared in several national publications. In May 1962, the group organized picketing on Wall Street, targeting the companies investing in Huntsville. Signs read: "Don't invest in Huntsville, Ala., it's bad business" and "To bring new plants and businesses to Huntsville aids segregation and subjects additional employees to racism." The next month, a group led by Sonnie Hereford III conducted a similar protest in Chicago. Hereford made a point to seek out representatives of the aerospace companies with a presence in Huntsville. The pressure worked. In the early days of the protests, Mayor Searcy agreed to a biracial committee to plan for desegregation, but he told the CSC that no white leaders would agree to serve. After the arrests and the public protests, city leaders caved. Two white businessmen joined the commission and, in discussion with black leaders, worked out a plan by which the city would desegregate. Hereford remembered that "once [the city] saw that desegregation was coming, it tried to make things go as smoothly as possible." Sheryll Cashin's analysis was blunter: "There was money to be made, and the race issue could not stand in the way of progress."[17]

The city's acquiescence came only after a determined and insistent campaign by African American residents; as Hereford explained, "It was because we'd kept up the pressure." And, as in other southern cities, the protestors were met with strident opposition and, on occasion, physical violence. In fact, a white NASA employee, Marshall Keith, who accompanied young black men to a Woolworth's, was abducted from his home at gunpoint, blindfolded, stripped, beaten, and sprayed with a chemical irritant. Hank Thomas had the seat of his car impregnated with an irritant as well, a chemical later found to be oil of mustard. Both Cashin and Hereford remembered receiving personal threats and verbal abuse in response to their work for civil rights.[18]

Yet observers still differentiated between events in Huntsville and the reception of civil rights protests in the rest of the state. This difference was most notable in the efforts to integrate the city's educational institutions. In June 1963, two black federal employees, Marvin Carroll and Dave McGlathery, brought suit against UAH for denying their applications. Carroll eventually withdrew, but McGlathery was added to the case of Vivian Foster and James Hood, who were seeking admission to the main Tuscaloosa campus. Governor George Wallace sought to halt integration. In addition to his infamous promise to "stand in the schoolhouse door," he threatened to contact Senators Lister Hill and John Sparkman in an effort to have McGlathery transferred, and he even suggested that such efforts at integration would prove detrimental to Marshall's future in Alabama. But while Wallace's stand in Tuscaloosa made him the face of civil rights resistance, in Huntsville civic leaders rejected the governor's attempt to prevent integration. Mayor Searcy told reporters that "we do not expect trouble" and that "law and order [would be] preserved." While the local National Guard mobilized as a precautionary measure, McGlathery "walked unescorted through a front door" of the extension center and registered for courses. Laurence Stern, writing in the *Washington Post*, remarked that "as far as the town [of Huntsville] is concerned he could walk onto the University campus with scarcely a ripple of trouble."[19]

The integration of the city's public schools followed a similar trajectory: again, black citizens pressed for change, and again, in the face of Wallace's threatening rhetoric, Huntsville quietly complied with integration orders. The effort was spearheaded by Sonnie Hereford III, who filed suit along with three other families in late 1962. The case, heard in Birmingham along with other cases from Birmingham, Mobile, and Tuskegee, resulted in a victory for integration. In September 1963, despite a brief temporary closure ordered by Wallace, Sonnie Hereford IV enrolled as a student at Fifth Avenue Elementary, accompanied by his father, a photographer, and a few plainclothes policemen. The city's police chief praised the integration as "smooth as silk." The *Washington Post* praised Huntsville as a "brighter" example of what was possible in a southern community, one in which desegregation of schools happened peacefully.[20]

Of course, "brighter" Huntsville had a very important mitigating factor that shaped the discussion around integration: the city's economy was uniquely sensitive to questions of federal funding, and in both cases of relatively uneventful desegregation, the presence of federal facilities shaped Huntsville's response. Seeking to integrate UAH, John and Joan Cashin specifically recruited McGlathery and Carroll because they were black engineers employed at Marshall, trying to enroll in technical courses offered to train aerospace workers. Despite state and local pressure, officials at the space center backed integration efforts. Similar pressures shaped Hereford's experience. The doctor later learned that his son's peaceful integration of the elementary school came thanks to a backroom deal between city leaders and Wallace allowing for only a small show of resistance; city leaders "didn't want to lose those federal contracts." Thus, Huntsville's "brighter example" was less a response to the city's "racial harmony" or support for civil rights than it was the threat that continued opposition might threaten the city's economic lifeblood. As the *Washington Post* noted, "Huntsville is not disposed to look too harshly on the Federal government."[21]

Federal officials understood this relationship, too, and they proved willing to use threats to force such accommodations. Under orders from the Department of Justice to tighten the screws on businesses working with Wallace in Alabama, NASA's James Webb contacted a number of aerospace companies, including Chrysler, General Electric, IBM, and Thiokol, who agreed to voice their concerns about the governor's actions. In late October 1964, Webb went further. Speaking to the Huntsville Industrial Expansion Committee, he announced that because of "social conditions in Alabama," the space agency might transfer some of its top scientists and executives from Marshall to the agency's Michoud Assembly Facility outside New Orleans, noting that the decision "will directly depend on our ability to improve our record of attracting senior executives from industry to the Huntsville installation." In fact, recruiters at Marshall routinely described difficulties in convincing educated engineers and technicians from other parts of the country to relocate to Alabama. Some supporters of civil rights even suggested boycotting states like Alabama, which continued to resist federal civil rights legislation. As the *Chicago Defender* asked, "Why can't someone in the government at least wonder out loud about the advisability of transferring Redstone to nearby Georgia or Tennessee?"[22]

City and state leaders scrambled to respond. Senator John Sparkman called the threat of a boycott "economic extremism" and reassured investors that the state was "bursting with opportunity for you and your business." His colleague, Lister Hill, echoed the common argument that economic development would eventually improve conditions in the state. The city's congressional delegation maintained personal contact with NASA and with the Democratic Party's leadership as a first line of defense against such threats. When NASA suggested the possibility of a move from Marshall, Sparkman and Representative Bob Jones personally contacted Webb and talked directly to President Lyndon Johnson, who privately promised the men that nothing would "disturb operations" in Huntsville.[23]

At the same time, NASA leaders worked within the policy recommendations of the Kennedy and Johnson administrations to encourage Huntsville and other southern centers to adhere to antidiscrimination and equal opportunity mandates. The agency's administrators were involved in committees to study and implement such mandates, and in the wake of the Civil Rights Act of 1964, Webb urged contractors to "use your good offices" to help with the "early, orderly and effective implementation" of desegregation. In Huntsville, one of the best examples of the impact of such pressure came with the formation of the Association of Huntsville Area Contractors (AHAC), a group created to help implement equal employment mandates by encouraging contractors to work with local institutions to encourage a more diverse white-collar workforce. Trying to address a legacy of discriminatory and segregationist educational opportunities, AHAC worked with Alabama A&M and Oakwood College in the hopes of creating engineering programs and pathways and encouraging college-educated blacks to apply for positions. The organization also pressured the city to improve housing options and desegregate downtown hotels in the hopes of attracting out-of-state minority applicants. Milton Cummings, an engineer, leading booster, and spokesman for AHAC, promised to "make our citizens more conscious of our responsibility in the area of housing, education, and the availability of private and public facilities."[24]

Webb told Marshall's chief, Wernher von Braun, to give the issue of civil rights his personal attention. The scientist was a perennial booster for the city's space and rocket industry, fighting for new programs, campaigning for federal funds, and working with contractors to help them bid for projects; he was sensitive to the impact that Alabama's resistance to civil rights might have on the success of his facility. Von Braun told city and state leaders to address racial inequalities, stating that "obstructionism and defiance . . . can hurt and are hurting Alabama." In a speech to the Huntsville–Madison County Chamber of Commerce, von Braun told his audience of the difficulty he and others faced in recruiting, thanks largely to the state's "public image," and advised them to make changes in order to "match the very keen competition of other cities of the country for the future space dollars." In particular, he noted the recent creation of the biracial committee, and he praised AHAC for participating in the space center's equal opportunity initiatives. Such actions would, he hoped, result in an improved image that might persuade national talent to come to the Rocket City "willingly."[25]

A few months later, speaking to the state chamber, von Braun more forcefully made his case for reshaping the state's image. He stressed the impact of federal funds on Huntsville and the state, noting that Redstone and Marshall employed nearly 22,000 people, conducted business in forty-two states, paid employees $215 million annually, and returned nearly $7 million in income tax alone. He noted that "our program has impacted heavily on Alabama's prosperity [and] our state should take an approach that is positive and constructive to solve the problems of this era . . . and we should work hard to shed the labels of obstructionism and defiance that have been applied to us." Given the "tremendous impact" that federal funds had on the state, he concluded, Alabama should be ready to follow the lead of other forward-minded cities and states and "accomodat[e] the forces of change" in order to preserve prosperity.[26]

This appeal for accommodation matched efforts by civic and business leaders in Huntsville and across the Sun Belt South. Worried about the impact of a negative "reputation" on economic growth and development, civic leaders accepted a modicum of desegregation and economic participation by African Americans. Efforts to comply with equal employment regulations, to encourage more black scientists and technicians at Marshall, to increase African American participation in course programs at UAH and Alabama A&M, and to desegregate public facilities across the city led observers to note the "bright" spot that was the Rocket City. Through 1965 and 1966, national correspondents routinely pointed to the city's anomalous position in Alabama: Huntsville put economics ahead of segregation, it was "spurning racism," and it was so far ahead of other southern towns that African Americans "tend to migrate to it."[27]

This narrative, that the efforts by moderate business leaders in the face of pressure by the federal government led to peaceful integration and racial cooperation, is too generous. Certainly, Huntsville avoided the outright conflict of Montgomery, Birmingham, and Selma. And ultimately, Huntsville's leaders proved more accommodating in the hopes of protecting its reputation and maintaining investment. But the Rocket City of the late 1960s was not the success story that many observers claimed. The same *New York Times* correspondent who reported that Huntsville had "spurned" racism quoted civil rights leader John Cashin that the city was simply "coasting on a liberal reputation." Black residents still found it difficult to buy houses or find high-paying jobs. The city was slow to expand basic municipal services to black neighborhoods, many of which remained mired in poverty and unemployment. One researcher specifically cited African American economic inequality as "grim evidence that not all of the citizens of the area have shared in the increasing prosperity of the space age."[28]

In fact, efforts to diversify the federal workforce resulted in minimal gains, and in the following decades the facilities struggled to meet equal opportunity guidelines. As the population of the city boomed from the 1950s to the 1970s, the black population declined, dropping from 32 percent to 12 percent, an indication that economic opportunities did not extend across the color line. As the historian Steven Moss notes, black technicians remained an "insignificant part of the Alabama workforce." In the 1970 census, which counted over 17,300 engineers and 2,300 computer scientists, only 197 and 33, respectively, were black. To be fair, such problems existed at other facilities in NASA's "Southern Crescent," including sites in Florida, Mississippi, Louisiana, and Texas. But taken as a whole, efforts to create a truly integrated workforce failed in the face of deep-seated inequalities. The legal restrictions that marked Jim Crow Huntsville fell thanks to the efforts of John and Joan Cashin, Sonnie

Hereford III, and hundreds of black men and women who highlighted the hypocrisy at work in the Rocket City, but the more insidious effects of social and economic discrimination remained long after public attention subsided.[29]

Under local and national pressure, Huntsville's leadership made a concerted effort to "accommodate the forces of change." Afraid to lose federal and private investment, civic leaders agreed to desegregate facilities and schools and work toward equal opportunities, but only after activists called attention to inequality and threatened the city's economic lifeline. Given the violence that accompanied civil rights struggles in other Alabama cities, Huntsville's relatively peaceful experience resulted in a reputation as the most "progressive" city in a state nationally known for George Wallace, massive resistance, and violent opposition to basic civil rights. Yet Huntsville's black leaders understood the shallowness of that characterization. As the nation's attention focused elsewhere, many of the underlying economic and social inequalities, difficult to address, unsuited to national press coverage, and thus nonthreatening to the city's business climate, remained in effect.

Notes

1. James E. Webb, untitled address, National Press Club, Washington, DC, May 15, 1965, in Box 299, Folder 15 (32, NASA, January–May 1965), Lister Hill Papers, W. S. Hoole Special Collections, University of Alabama (hereafter cited as Hoole).

2. Wernher von Braun, "Alabama's Future with the Space Program," June 1965, in Box 1, Folder "Dr. von Braun's Speeches 1965," NASA Files, National Archives and Records Administration, Southeast Branch, Morrow, GA (hereafter cited as NARASE).

3. Robert J. Norrell, *Reaping the Whirlwind: The Civil Rights Movement in Tuskegee* (New York: Alfred A. Knopf, 1985); Adam Fairclough, *Race and Democracy: The Civil Rights Struggle in Louisiana, 1915–1972* (Athens: University of Georgia Press, 1995); Stephen G. N. Tuck, *Beyond Atlanta: The Struggle for Racial Equality in Georgia, 1940–1980* (Athens: University of Georgia Press, 2001). John Dittmer's *Local People: The Struggle for Civil Rights in Mississippi* (Urbana: University of Illinois Press, 1995) combines the community study and the grand narrative of the movement in a particularly effective way.

4. J. Mills Thornton, *Dividing Lines: Municipal Politics and the Struggle for Civil Rights in Montgomery, Birmingham, and Selma* (Tuscaloosa: University of Alabama Press, 2002); Glenn Eskew, *But for Birmingham: The Local and National Movements in the Civil Rights Struggle* (Chapel Hill: University of North Carolina Press, 1997).

5. There is a broad historiography of white opposition to civil rights, nicknamed "massive resistance." See Numan V. Bartley, *The Rise of Massive Resistance: Race and Politics in the South during the 1950s* (Baton Rouge: Louisiana State University Press,

1969); Neil McMillen, *The Citizens' Council: Organized Resistance to the Second Reconstruction, 1954–64* (Urbana: University of Illinois Press, 1971); Dan Carter, *The Politics of Rage: George Wallace, the Origins of the New Conservatism, and the Transformation of American Politics* (Baton Rouge: Louisiana State University Press, 1995); Jeff Frederick, *Stand Up for Alabama: Governor George Wallace* (Tuscaloosa: University of Alabama Press, 2007); Chris Myers Asch, *The Senator and the Sharecropper: The Freedom Struggles of James O. Eastland and Fannie Lou Hamer* (New York: New Press, 2008). Recent works have complicated the image of resistance, most notably Jason Sokol, *There Goes My Everything: White Southerners in the Age of Civil Rights, 1945–1975* (New York: Vintage, 2006); Matthew D. Lassiter, *The Silent Majority: Suburban Politics in the Sunbelt South* (Princeton, NJ: Princeton University Press, 2006); Kevin M. Kruse, *White Flight: Atlanta and the Making of Modern Conservatism* (Princeton, NJ: Princeton University Press, 2005).

6. Elizabeth Jacoway and David R. Colburn, eds., *Southern Businessmen and Desegregation* (Baton Rouge: Louisiana State University Press, 1982); Benjamin Houston, *The Nashville Way: Racial Etiquette and the Struggle for Social Justice in a Southern City* (Athens: University of Georgia Press, 2012); Tracy E. K'Meyer, *Civil Rights in the Gateway to the South: Louisville, Kentucky, 1945–1980* (Lexington: University Press of Kentucky, 2010).

7. "The Need for More Industry," *Huntsville Times*, July 2, 1956, 4; "'As Redstone Grows . . . ,'" *Huntsville Times*, November 21, 1956, 4. See also Matthew L. Downs, *Transforming the South: Federal Development in the Tennessee Valley, 1915–1960* (Baton Rouge: Louisiana State University Press, 2014).

8. John Sparkman, press release, December 4, 1958, Box 303, Folder 4, Dept. of the Army, John Sparkman Senate Papers, Hoole; "Compromise to Profit All," *Huntsville Times*, December 4, 1958, 4; Downs, *Transforming the South*, 227.

9. Huntsville Industrial Expansion Committee, "Huntsville Has What It Takes," 41, 44, 50, Hoole; Sonnie Wellington Hereford III and Jack D. Ellis, *Beside the Troubled Waters: A Black Doctor Remembers Life, Medicine, and Civil Rights in an Alabama Town* (Tuscaloosa: University of Alabama Press, 2011), 90; William H. Chafe, *Civilities and Civil Rights: Greensboro, North Carolina, and the Black Struggle for Freedom* (New York: Oxford University Press, 1980), 6–8.

10. Hereford and Ellis, *Beside the Troubled Waters*, 18–27, quote on 26; Sheryl Cashin, *The Agitator's Daughter: A Memoir of Four Generations of One Extraordinary African-American Family* (New York: PublicAffairs, 2008), 93–95; Chafe, *Civilities and Civil Rights*, 9–10.

11. Hereford and Ellis, *Beside the Troubled Waters*, 75, 88–89; Shelby Johnson to Sparkman, November 30, 1941, and Sparkman to Johnson, December 9, 1941, Box 34, Folder "Appointments: CWA, Letters Not Filed (1)," John Sparkman House Papers, Hoole; "Redstone to Let 200 Employees Go," *Huntsville Times*, June 13, 1945, 1. Monique Laney, who studied the sociocultural impact of the arrival of the German rocket scientists, noted that they did not immediately feel a sense of "solidarity" with black residents, even though individually, some treated black men and women more equally than their American-born neighbors. See Monique Laney, *German Rocketeers in the Heart of Dixie: Making Sense*

of the Nazi Past during the Civil Rights Era (New Haven, CT: Yale University Press, 2015), 133–134.

12. Enoc P. Waters, "Jim Crow Still in Business . . . but Slipping," *Chicago Defender*, August 18, 1945, 11; Hereford and Ellis, *Beside the Troubled Waters*, 55–57, 88–89; Laney, *German Rocketeers*, 126, 138–139.

13. Chafe notes that in Greensboro, the protests "circumvented" the "fraudulent communication and self-deception" that made up the environment of civility underlying the city's "progressive mystique." The protests in Huntsville had a similar effect. See Chafe, *Civilities and Civil Rights*, 138–139.

14. Mills Thornton notes that Montgomery lost a number of investments in the wake of the city's hostility to the bus boycotts. See Thornton, *Dividing Lines*, 320–22; James C. Cobb, *Industrialization and Southern Society, 1877–1984* (Lexington: University Press of Kentucky, 1984), 110; Smyer quote from Eskew, *But for Birmingham*, 279.

15. James Cobb, "Yesterday's Liberalism: Business Boosters and Civil Rights in Augusta, Georgia," in Jacoway and Colburn, *Southern Businessmen and Desegregation*, 151–153, 155–157; William Brophy, "Active Acceptance–Active Containment," in Jacoway and Colburn, *Southern Businessmen and Desegregation*, 139–148.

16. Holly Fisher, "Oakwood College Students' Quest for Social Justice before and during the Civil Rights Era," *Journal of African American History* 88, no. 2 (Spring 2003): 114–116; "Governor of Alabama Predicts Racial Strife," *Washington Post*, November 11, 1959, B14; "A Time for Caution and Wisdom," *Huntsville Times*, November 11, 1959, 4; Cashin, *Agitator's Daughter*, 130–131, 131–132, 140–141; Hereford and Ellis, *Beside the Troubled Waters*, 101–103, 110, 161n20.

17. Cashin, *Agitator's Daughter*, 143–148; Hereford and Ellis, *Beside the Troubled Waters*, 104, 108–110. Cashin and Hereford disagree on the names of the white members. Hereford writes that the two were Harry M. Rhett Jr. and William Halsey. Cashin names Rhett and James Johnston.

18. Richard Paul, "How NASA Joined the Civil Rights Revolution," *Smithsonian Air and Space Magazine* (March 2014), http://www.airspacemag.com/history-of-flight/how-nasa-joined-civil-rights-revolution-180949497/?all; "Integrationist Is Hurt," *New York Times*, January 22, 1962, 14; Cashin, *Agitator's Daughter*, 137–139; Hereford and Ellis, *Beside the Troubled Waters*, 91–92, 119.

19. Laurence Stern, "Policies of Gov. Wallace Unpopular in Huntsville," *Washington Post*, June 10, 1963, A1; "Third Student Admitted to Alabama U. at Huntsville," *Chicago Defender*, June 15, 1963, 1; Cashin, *Agitator's Daughter*, 150–151; Steven L. Moss, "NASA and Racial Equality in the South, 1961–1968" (MA thesis, Texas Tech University, 1997), 97–98.

20. "In Brighter Alabama," *Washington Post*, June 16, 1963, E6; "Huntsville, Ala., Schools Admit 10 More Negroes," *Washington Post*, January 28, 1964, A2; Hereford and Ellis, *Beside the Troubled Waters*, 115–119.

21. Claude Sitton, "Racial Test Due over Alabama U.," *New York Times*, March 25, 1963, 7; Stern, "Policies of Gov. Wallace"; Bill Austin, "U.S. Sues to Integrate City-County Schools for Army's Children," *Huntsville Times*, January 18, 1963, 1, 2; "4 Areas in South Are Sued to End Pupil Separation," *New York Times*, January 19, 1963, 1; "Huntsville, Ala.,

Schools Admit 10 More Negroes," *Washington Post*, January 28, 1964, A2; Cashin, *Agitator's Daughter*, 150; Hereford and Ellis, *Beside the Troubled Waters*, 118–119.

22. John W. Finney, "NASA May Leave Its Alabama Base," *New York Times*, October 24, 1964, 12; Lillian S. Calhoun, "Confetti," *Chicago Daily Defender*, October 24, 1964, 4; Michael J. Neufeld, *Von Braun: Dreamer of Space, Engineer of War* (New York: Alfred A. Knopf, 2007), 395–396; "Help Offset Alabama Boycott, Sparkman Tells Industry Group," April 28, 1965, Box 8, Folder "CR 4/28/65," Sparkman Senate Papers, Hoole; Downs, *Transforming the South*, 235–236; Moss, "NASA and Racial Equality," 98–99, 104.

23. Form letter, Sparkman, April 23, 1965, Folder 8: "CR 4/23/65," Sparkman Senate Papers, Hoole; "Help Offset Alabama Boycott, Sparkman Tells Industry Group," April 28, 1965, Box 8, Folder "CR 4/28/65," Sparkman Senate Papers, Hoole; Hill to Louise Reed, January 7, 1965, Box 188, Folder "212 (4-B, Military Installations, RSA, January 1965–December 28, 1966)," Lister Hill Papers, Hoole; Sparkman to Frank W. Boykin, October 27, 1964, Box 66A677-7, Folder "Fed. Gov't, NASA," Sparkman Senate Papers, Hoole.

24. Moss, "NASA and Racial Equality," 39–40, 48–49, 101–102, quote on 49; Laney, *German Rocketeers*, 59, 125–127, quote on 126; Andrew J. Dunar and Stephen P. Waring, *Power to Explore: A History of Marshall Space Flight Center, 1960–1990* (Washington, DC: NASA History Office, 1999), 117–119, Brown quote on 119. Moss argues that NASA balanced such mandates with the necessity of working alongside southern communities and navigating segregated society. On the whole, though, he suggests that NASA did as much as it could, and as much as could be expected of a federal agency, to advance the cause of civil rights; see Moss, "NASA and Racial Equality," 63–64.

25. Neufeld, *Von Braun*, 385–386, 395–396; Wernher von Braun, "Huntsville in the Space Age," December 8, 1964, in Box 1, "Dr. von Braun's Speeches 1964," NASA Files, NARASE; Dunar and Waring, *Power to Explore*, 118–119, 123.

26. Wernher von Braun, "Alabama's Future with the Space Program," June 1965, in Box 1, "Dr. von Braun's Speeches 1965," NASA Files, NARASE.

27. Paul, "How NASA Joined the Civil Rights Revolution"; Cobb, "Yesterday's Liberalism," 155–157; Edward C. Burks, "Huntsville, Ala., Spurning Racism," *New York Times*, May 23, 1965, 56; Ruby Abramson, "Huntsville Is on Moon Boom," *Washington Post*, June 9, 1966, F13.

28. Burks, "Huntsville, Ala.," 56; Mike Hollis, "New Year Prospects Not Bright for Some Huntsvillians," *Huntsville Times*, January 1, 1970, 4; Thomas Franklin Morring, "The Impact of Space Age Spending on the Economy of Huntsville, Alabama" (MA thesis, MIT, 1964), 62–63.

29. Dunar and Waring, *Power to Explore*, 124; Laney, *German Rocketeers*, 51–52; Moss, "NASA and Racial Equality," 82, 85, 94, 138.

6

NASA, the Association of Huntsville Area Contractors, and Equal Employment Opportunity in the Rocket City, 1963–1965

BRIAN C. ODOM

On August 26, 1961, Huntsville resident Marshall Keith wrote to John Wilcock, editor of the New York–based magazine the *Village Hippie*.[1] Keith stated that the "Old Guard Huntsvillians" were continuing to fight a rear-guard action against desegregation in the area, but that change was in the air. The dynamic growth of defense spending on the army and funding for the space program had made Huntsville dependent "almost entirely on income produced by the arsenal," making Huntsville the "logical place to begin integration in Alabama." Area civil rights activists, through boycotts and sit-ins, had already proven that the merchant's "purse strings control his social conscious." Keith appealed to Wilcock to "tell the NAACP [National Association for the Advancement of Colored People] that Huntsville is ripe."[2]

Between 1961 and 1965, the federal government acted through the National Aeronautics and Space Administration's (NASA) Marshall Space Flight Center to promote a program of equal employment opportunity. Marshall enacted this program in its own facilities and those of its private industry contractors, coordinating these policies through both funded and volunteer community groups. The promise of government contracts worth billions of dollars caused many to worry that the politics of Montgomery, Alabama, and the violence in Birmingham and Selma, Alabama, would tar Huntsville by association and undermine the financial foundations of the Rocket City. As the decade progressed, NASA's budget increased from $500 million in 1960 to $5.2 billion by 1965, roughly 5.3 percent of the federal budget.[3] In order to ensure compliance with federal directives, Marshall established a compliance program focused primarily on recruiting black engineering students from Historically Black Colleges and Universities (HBCUs) for its cooperative programs and developing local training programs to train area blacks to qualify for menial and clerical jobs at the center. As the decade wore on and federal programs shifted from civil rights legislation to the War on Poverty, Marshall management paid less attention to developing black engineers and embraced temporary opportunities for local poor blacks and whites. The celebrated achievement of Apollo 11 in July 1969 overshadowed the center's paltry performance of increasing the overall number of black engineers and a minimal engagement with community programs designed to equip blacks for Space Age jobs.

While the results of Marshall's efforts at equal employment produced minimal results, coverage of its program is critical to understanding how funding for the space program impacted the revolutionary transformations that took place in Huntsville's institutions of black education and how local civil rights leaders utilized that funding as leverage against city officials and area businesses to overturn Jim Crow segregation. This chapter surveys the development of equal employment at Marshall and assesses the minimal gains made at the center by the close of the decade. While the efforts at the center paled in comparison to events beyond the gates of Redstone Arsenal, they serve as a critical starting point for understanding black engagement with Space Age employment opportunities.

Improving the image of the southern communities that made up the Space Crescent and effecting real change in hiring practices would be difficult tasks.[4] President John F. Kennedy's Executive Order 10925, transmitted on March 6, 1961, underscored the duty of the federal government to "promote and ensure equal opportunity for all qualified persons, without regard to race, creed, color, or national origin," in the employment practices of both government agencies and their contractor partners. The order also created a body known as the President's Committee on Equal Employment Opportunities (PCEEO), which would replace Eisenhower's Committee on Government Contracts and the Committee on Government Employment Policy. This body scrutinized and studied employment practices in the federal agencies and developed and recommended concrete steps for affirmative action to ensure the success of the program.[5]

NASA leadership understood how the location of much of the space program's facilities in the Jim Crow South was problematic. In an April 15, 1963, briefing to the NASA administrator, NASA's principal compliance officer, Alfred Hodgson, pointed out the many problems experienced by NASA centers in the implementation of Kennedy's Executive Order 10925. According to Hodgson's 1963 report, Marshall, Houston, and the Launch Operations Center in Florida all represented problematic areas in need of immediate attention. Hodgson underscored that blacks constituted approximately 11 percent of the total US population and were the "largest ethnic group against which there has been a widespread, discernible economic and social prejudice." Hodgson argued that it was because of this that the PCEEO had called upon federal offices to bring their considerable resources to bear on alleviating the situation as much as possible.[6]

Located in the heart of the Deep South, Huntsville epitomized the problematic aspects of Hodgson's concern. Tasked with delivering Marshall's compliance report to NASA headquarters, Paul L. Styles brought years of experience to the job of working in labor relations. A native of Knoxville, Tennessee, Styles had spent his school years outside the South writing for the *Cleveland Plain Dealer* and Chicago's *Voice of Labor* newspapers before coming to Huntsville to work as a weaver, loom fixer, and wrap roller at Lincoln Mills and writing for the *Huntsville Times* from 1928 to 1937. In 1937, Styles left Huntsville to take a position as field examiner with the National Labor Relations Board (NLRB) in Atlanta, and in 1944 he became director of disputes of the Atlanta Regional War Labor Board, where he negotiated labor disputes in the southeastern region during World War II. Styles became regional director of the Atlanta NLRB in 1950 and, for a time during the Truman administration, was part of the NLRB in Washington, DC. Styles came to Marshall in 1961 to become chief of the Marshall Center Industrial Relations Office. His travels outside the South and work with the NLRB shaped Styles into someone who understood the problems facing the industry in terms of both civil rights compliance and labor contracting. Throughout his career, Styles corresponded with several high-profile individuals from both sides of the ideological spectrum, including labor leaders George Meany and Phillip Murray, Supreme Court justice Hugo Black, and US senator Lister Hill. Known for his affinity for his "well-crusted, hook-shaped pipe," Styles once remarked that "labor relations in the space age haven't changed much from the horse and buggy days."[7]

The statistics presented by Styles in April 1963 on behalf of the Marshall workforce to the administrator's briefing were representative of the problem areas Hodgson had mentioned in his presentation. Blacks represented only 0.1 percent of the total workforce at the management level (GS 12–GS 18), with no salaried blacks in the wage positions making over $8,000 per year. Styles observed that more than anything else, the location of the Marshall Center in the Jim Crow South had adversely affected its efforts to hire more qualified black employees. Styles argued that in the Deep South, "segregation and discrimination have been a way of life for one hundred and fifty-four years," and a "great deal of work and particularly education" would be necessary before the center made any real progress regarding discrimination because of race.[8] Whether this admission of the stark realities the center faced was intended as a starting point for reform or an excuse for poor performance is debatable.

Industrial firms also understood that the state's negative image was bad for business. Racist, exclusionary policies were already beginning to result in lost investments and missed opportunities. Bobby Wilson argues that an awareness of the dangers posed by civil rights activities to a city's image was evident as early as the 1955 Montgomery bus boycott. Because of its own negative image in regard to race relations, Montgomery had missed an opportunity to secure a DuPont facility and four other plants to competing cities.[9] In Huntsville, the Association of Huntsville Area Contractors (AHAC) served as a focal point of the ongoing discourse among Marshall, local industrial contractors, and the Huntsville black communities. The association was comprised of representatives from aerospace firms like Lockheed, Haynes International, and IBM, and L. C. McMillan from Prairie View A&M University occupied the executive director position. AHAC served as both the catalyst and critical liaison between federal and community efforts that cemented the gains made by Huntsville's civil rights activists and community leaders over the previous years, and it provided a mechanism for an innovative approach to educational initiatives and the coordination of new federal hiring guidelines.

The development of AHAC was critical to integrating equal employment efforts in Huntsville both with civil rights activities and improving local black educational institutions. The spark for creating the group was an April 19, 1963, letter from Webb to then Marshall Center director, Wernher von Braun, highlighting the center's struggles implementing an effective equal employment opportunity program. Webb pointed out the importance of the policy "for the good of the country" but was understanding of how the policy might be "time consuming, difficult to implement, and full of intangibles." The administrator mentioned how NASA efforts over the past year pleased both the vice president and the chair of the Civil Service Commission. Webb noted that the problems von Braun faced were more difficult than those that existed in other parts of the country.[10] Webb understood the degree of difficulty faced by the center but was unwilling to offer any exemptions.

In his response to Webb, von Braun pointed out the steps Marshall was taking to comply with the guidelines, including activities the center was taking in the black community to identify qualified black job applicants. Von Braun stated his belief that in time, local black institutions such as Oakwood and Alabama A&M Universities would begin to produce "qualified applicants." However, the center director pointed to larger related issues that were brewing in both Huntsville and Tuscaloosa at the University of Alabama. Von Braun related to Webb that the trouble surrounding the enrollment of three black students at the University of Alabama (including one Marshall employee) constituted a "severe obstacle in attracting qualified Negro applicants."[11] In the summer of 1963, it was increasingly difficult for NASA and Marshall to remain insulated from the tumultuous events taking place in Tuscaloosa and Birmingham.

The space agency needed the help of organizations in the surrounding communities capable of both keeping a safe distance from federal entanglements and bringing economic and political pressure to bear on local and state leaders. AHAC proved capable of providing both. During its initial meeting in the offices of Brown Engineering in Huntsville on July 5, 1963, AHAC appointed Milton Cummings as its temporary spokesperson and unanimously approved points of agreement that included increasing minority hires in accordance with Executive Order 10925. The representatives agreed to "aggressively seek out Negro and other minority people throughout the United States," place black employees in personnel departments to ensure compliance, cooperate with local black educational institutions, and use their substantial influence in the Huntsville community and across the state to "make our citizens more conscious of our responsibility" and "help the colored high schools achieve equality with the white high schools" in both facilities and instruction. When members pointed out their companies received few applications from prospective blacks, AHAC representatives agreed that this reflected "a weakness in Negro educational opportunities," something the association would have to address going forward. There was an agreement to provide reports of their activities from time to time to NASA, inviting any "suggestions for improvement."[12]

Members of the black community took note of the association's activities. Clyde Foster, a longtime African American employee at Marshall, remembered AHAC as "the coming together of all the pillars of the community." According to Foster, the purpose of the organization was to respond to the racial problems in the region and figure out what could be done with the federal funding in efforts to combat the dominant image of George Wallace. To Foster, the group's activities made desegregation efforts in Huntsville "much easier than anywhere else . . . not only in Alabama, but in the South."[13] In this way, AHAC and its member institutions served as a clearinghouse for legislation related to civil rights and served as a model for institutional reform.

Marshall management understood the need to demonstrate its own commitment to the principle of equal employment. One of the first programs to make an impact was a Marshall Personnel Office program dedicated to recruiting qualified black graduates to accept jobs at the center—a program working in tandem with a university cooperative student initiative with HBCUs with engineering programs. In the fall semester of 1963, six black students from Southern University in Baton Rouge, Louisiana, arrived at Marshall. Warren J. August, Morris Pipkins Jr., Lawrence Champagne, Howard Turnley, Hugh V. McKnight, and William Porche were electrical engineering majors who spent their time at the center working in electronics development and quality checkout of the Saturn launch vehicle. These programs would continue to grow throughout the 1960s, expanding to embrace other HBCUs, including Xavier and the Tuskegee Institute.

A letter from Taylor to the adviser for National Capital Affairs, Charles A. Horsky, offers insight into how local economic and political groups developed and utilized these connections. While debating the best strategy for developing equal opportunity housing for federal workers, Taylor brought up his group's work in Huntsville, pointing out that most of the residents of the city were dependent upon the contracts associated with NASA. Taylor stated that the PCEEO had used this fact to insist that "the Government agencies and the Government contractors cooperate to bring about gradual desegregation in the town itself, of recruitment and training offered non-whites in the locality," and in public and private educational institutions.[14]

Taylor underscored how this was achievable, saying, "We believe that the Government contractor can reach the private builders through the local banking and financial institutions," and in this way, "it will be possible to establish a different housing pattern in Huntsville." What worked in Huntsville "could certainly be done in other places." According to Taylor, the key to the process was to be "firm, but reasonable," and what he referred to as "locking the barn *before* the horse is stolen." Taylor argued for the introduction of these terms before contracts were finalized, something that would give additional leverage to contracting agencies in their dealings with local politicians, bankers, and builders. Working this way promised to create an environment "conducive to the success of the effort" and enable pressure that would permit the contracting agency to "get a plan and not mere promises."[15]

Interaction between federal and local groups continued through the fall of 1963. A meeting took place on November 21, 1963, between Milton Cummings, Al Hodgson, Marion Kent, Hobart Taylor of the PCEEO, two army representatives, and the AHAC's newly appointed executive director, L. C. McMillan. McMillan, an African American, had previously served as an administrator for Prairie View A&M College in Texas.[16] He joined the organization one month prior, on November 1, remembering later how the group impressed him by attempting to "slice away the fat and get right down to the meat of the problem."[17] By the end of the year, AHAC consisted of more than twenty firms with a combined workforce of more than 12,000 employees.[18]

Marshall continued to emphasize its cooperative programs with regional black colleges. On March 14, 1964, Marshall administrator Harry Gorman recounted to Alfred Hodgson the center's actions over the past two years, including an Administrative Intern Program and the cooperative program with Xavier and Southern Universities that would grow to eight by the end of the summer. Gorman argued that Marshall had made a difference in the Huntsville community, making it "one of the most objective and progressive in the South." Gorman pointed to AHAC as a "prime example" and an organization that had made a "tremendous contribution in affording opportunities to Negroes with technical, professional, and administrative skills." But more important, Gorman pointed out that as a matter of policy, Marshall would no longer be "officially represented at events which may be of beneficial interest to NASA but as a matter of policy exclude Negro attendance."[19] Lee White reiterated this message in a memorandum to the heads of departments and agencies on June 12, 1964, asking that all addressees take steps to "ensure that this position is understood throughout the Federal Government."[20]

One of Marshall's first actions in 1963 was to hire someone capable of recruiting African Americans to work at the center. The person selected for the job was Charles Smoot. Smoot began taking recruiting trips right away, traveling to black engineering schools across the country with strong math and science programs. Smoot remembered that there were "more jobs than you had people that could do them." This problem stood at the core of Marshall's efforts to implement an effective equal employment opportunity program. With so few qualified black engineers graduating every year and so many firms scrambling for their services, convincing recent graduates to come to the South to work for the federal government for less money than they could earn in the private sector in places like Southern California was certainly problematic.

As the decade wore on Huntsville's reputation improved in terms of segregation, but the state of Alabama's overall negative images in terms of race relations remained a major challenge to recruiting black graduates. Smoot noted that while the city of Huntsville was "better" than Birmingham in terms of race relations, it was "not significantly better." For Smoot, there was "no place like Birmingham in terms of segregation; that was a terrible time." He believed that the federal funding had "a lot to do with the openness of the town," which had no notorious individuals like Bull Connor, but rather had people like Milton Cummings, who, Smoot believed, understood the "need of money coming here" and convinced other white elites that "whether you like it or not, you are going to have to do certain things" if they wanted to keep that money flowing to the city. According to Smoot, it was the funding for NASA that "made this city and the fathers of this city do things sooner than they would have on their own." If they wanted to keep the "federal dollars coming in," keep the Army Missile Command, and encourage "people like Wernher von Braun and a group to come and live in a community where there's peace and tranquility, there are certain things you can do and certain things you can't do." It was clear to Smoot that "this group of people wanted this, so they did this in order to have that."[21]

Marshall management presented Smoot's efforts to Hobart Taylor in November 1964 as one of high points of the program. In this conversation, center management characterized Smoot as an employee who showed "promise in developing into an excellent recruiter for predominately Negro colleges and institutions" as well as in representing the center at other open house and career day programs.[22] The success of Marshall's ability to recruit talent from outside of Alabama was critical to the effective implementation of an Equal Employment Opportunity (EEO) program at Marshall. African American institutions required more resources from the city to fill the positions.

By the late spring of 1964, AHAC coordination and the expansive grassroots efforts of the Huntsville civil rights groups were beginning to pay dividends in terms of eliminating continued aspects of segregation in the city. On May 9, Macaluso provided Taylor with an overall update on Huntsville, reporting that the city had integrated a fifth of local motels as well as a quarter of the restaurants and drive-ins, the public golf course, one Scout troop, and over half of the lunch counters. Additionally, the "leading bank" was training six black tellers, and the three largest department stores were employing black salespeople. In the local government, the previously all-white country road crew now included ten blacks, while those serving on the city police force had grown to three, with another two being "actively recruited." The Management Training Center was now providing courses in the areas of "mechanical, clerical, and technical skills for integrated classes" for sixty-two people, eleven of whom were black.[23]

Macaluso's report also conveyed problem areas and remaining holdouts. Movie theaters remained segregated, along with the bowling allies, private schools, and the YMCA. The housing issue remained uncertain, and there was no indication of the rate of progress in integrating public accommodations. Macaluso's report also mentioned a letter from Milton Cummings that had made the rounds to Marion Kent at Marshall and McMillan. Cummings's letter indicated the group's expectation that a black member of the Bi-Racial Committee serve as the liaison between the black police and the city council. Cummings also noted that many locals, including the editor of the *Huntsville Times*; the director of the First National Bank, Harry Rhett; and Will Halsey were all out in the Huntsville business community persuading people to comply with de-segregation.[24] Macaluso noted that there had been some hostility toward Cummings from his fellow AHAC members who felt he was "running a one-man show." But Macaluso hoped it would be possible to "channelize" the hostility of the AHAC members into making them more active in the association's activities.[25]

The passage of the Civil Rights Act of 1964 on July 2 represented a major turning point in Marshall's and NASA's overall approach to equal employment. NASA administrator James Webb wrote to von Braun on July 13 to reinforce the urgency of "vigorous and effective leadership to recognize that the Civil Rights Act is now the law of the land." Webb told von Braun that because of Marshall's position in the Huntsville area, his personal leadership "fostering constructive attitudes" would serve a criti-cal role in the implementation of the new legislation. He urged von Braun to persuade everyone in leadership positions at the center to promote voluntary compliance with the law.[26] Webb understood that the political climate in Alabama was unfavorable to the legislation but also recognized that the economic importance of the center could ease acceptance in the region. Webb wrote a similar letter to the leadership of aerospace asso-ciations, including Aerospace Industries of America and the American Institute of Aeronautics and Astronautics, urging them to act quickly and firmly in supplying the leadership necessary in the implementation of the Civil Rights Act.[27]

Webb wrote a letter the same day to the associate special counsel to the president, Lee White, spelling out the space agency's intentions toward implementation of the new legislation. Webb stated that NASA was draft-ing regulations to preclude any discrimination in administering grants made available through NASA. All key NASA officials received a memo with Webb's signature to serve as guidance to federal and local leadership. Webb also planned to issue a directive concerning "possible acts of intimi-dation arising in relation to the civil rights movement."[28] Marshall needed to demonstrate its willingness to comply with the law and demonstrate what actions the center was taking to effect positive change. At Marshall,

Marion Kent and Dave Newby drafted a letter to von Braun reiterating many of Webb's points and called upon the recipients to offer their "assistance in fostering constructive attitudes" in the "effective and orderly implementation of the law." The letter asked these individuals to "supply the leadership which the President feels is so essential" to the success of the law.[29]

Von Braun's response to Webb's letter on August 11 outlined the actions Marshall had taken on the Civil Rights Act to that point. Von Braun noted a letter written on his behalf to Marshall employees urging their "prompt and willing" support of the act and asked all firms holding contracts greater than $25,000 to "exercise their leadership" with employees and acquaintances and "to foster constructive attitudes" necessary to implement the act. Von Braun also understood the value of community engagement. The center director mentioned meetings with members of the Marshall Advisory Committee as well as with the local Community Relations Committee (Bi-Racial) responsible for coordinating the program to comply with the Civil Rights Act. Management also scheduled meetings with contractors at Marshall's newly operational Mississippi Test Operations in Hancock County that would include the "leading citizens" in the area and additional government agency representatives.[30] Only stakeholder buy-in would produce normative change in Huntsville.

Sometime that July, Hobart Taylor reported to President Johnson about the progress made by the PCEEO. Topping the list was Huntsville's cooperative effort to desegregate several aspects of community life. Taylor stated that this action had produced an overall environment in which "all residents can lead more productive and useful lives." He noted a "basic change in attitude on the part of the managers of American industry" and that for the "first time" industry leaders had "undertaken a rigorous self-scrutiny to determine their true attitudes." This process was leading them to throw off the "binding and restricting residue of the past" and to begin an "examination of new human resources . . . both from humanitarian and practical motives." Taylor observed a "growing awareness of the moral, economic, and social cost of discrimination." However, the situation was in no way concluded. Much work remained because, in Taylor's mind, there could be "no true equality of opportunity in an unequal society."[31]

Leadership at both the community and federal levels recognized the importance of a good relationship with civil rights leaders. AHAC executive director L. C. McMillan wrote to Hobart Taylor on November 9, 1964, informing him of Hereford's telegram and passing along a copy that the author had given him personally. McMillan understood the implications of the letter, hinting to Taylor that he should respond to this kind of uneasiness during his upcoming visit to Huntsville. McMillan expressed his concern to Taylor that both NASA and army leadership, while committed to the EEO program, were having trouble showing any "tangible evidence to the Negro public." The leadership did not "communicate too well with the Negro community," something McMillan sought to alleviate by getting some "top flight Negroes on their EEO staffs."[32]

McMillan's criticism went well beyond any simple failure to communicate with the black community in Huntsville. The AHAC executive director accused NASA and army EEO leaders of making excuses and passing blame. McMillan claimed that this group represented part of the problem, publicly supporting equal employment while undermining it in private conversations. In McMillan's opinion, this duplicity prevented any lasting affirmative action program from taking root. The problems in Huntsville were difficult. McMillan pointed out that housing was a continued source of concern for both Marshall recruiters and Huntsville's black community, with plans for development striking many black leaders as an "effort to continue a 'ghetto' layout." McMillan thought local boards and committees should appoint more black members, while efforts to help Alabama A&M and Oakwood Universities should move beyond mere token efforts. For McMillan, the city needed to be "challenged to fulfill its destiny." McMillan told Taylor, "All the ingredients are here. We need the flavor of men of good will facing up to their responsibilities."[33]

The passage of the Voting Rights Act in 1965 and the shift to broader reforms targeting poverty across the nation signaled a dramatic change in federal funding priorities toward President Johnson's "Great Society" programs. By shifting the discourse from race to class, the move from equal employment to economic opportunity allowed Marshall leadership to directly address the center's position on Huntsville's disadvantaged populations. At a May 13, 1966, meeting of the Alabama chapter of the International Association of Public Employment Services, von Braun delivered a speech addressing the recent federal legislation intended to assist the "unfortunate, helpless and hopeless" of the country. The language of the speech focused upon the "humanitarianism" of the "War on Poverty" program and how state and local leadership should embrace efforts to utilize "all our manpower to its fullest extent, including both Negro and white." Von Braun pointed to necessary improvements in education without mentioning black institutions, but he did address the continued call to be "more sensitive" to equal opportunity and fair employment practices. In a final marked-up copy of the speech filed in the archives, a portion of the original speech addressed how Arkansas had "reacted violently" to school integration and "fought all over again an issue that had already been lost." Another section removed discussed how the government of Alabama had done "something positive about the plight of the minority Negro groups" and asked if the people of Alabama had learned "anything from Arkansas' bitter lesson." Speechwriters removed these sections and replaced them with another calling upon Alabamians to "take the lead in the forward march of humanitarianism" that was then "sweeping our nation and the world."[34] This change reveals an important aspect of what could and could not be said in the public sphere.

The end of the decade brought several critical evaluations of Marshall's EEO program. In May 1971, the director of the Manpower Office at Marshall, Paul Styles, wrote to now center director Eberhard Rees, informing him of the situation in terms of black employee numbers at the center. In a time of an overall reduction in force and a "no hire" situation coming on the end of the development phase of the Apollo program, the overall picture remained bleak. Between 1963 and 1969, the center's College Relations Program resulted in the hiring of only thirteen black engineers, something Styles argued was "quite significant" when considering the overall availability in the field. Styles noted Department of Labor statistics from a report by Robert Kiehl that "only two percent of all engineering students are black and about half attend six predominately black universities." The Kiehl report also stated that over the past eight years there had been "virtually no increase in the number of black engineering students" aside from a "few special programs."[35]

The numbers beyond engineering were not much better. Working with limited information, Styles determined that of 170 clerk-typists hired in 1968, only 21 were black—the total number available from the Civil Service Board of North Alabama registers. Having previously prohibited the collecting of minority group job numbers, the Civil Service Commission changed its position on September 30, 1969. The numbers reveal a dismal performance in terms of minority hires. A total of seventy-two black employees out of a total of 5,800 employees at the Marshall Space Flight

Center represented 1.24 percent of the overall workforce. The number of black workers employed by support contractors at the center was somewhat better, with a total of 249 minorities out of a total of 3,435 employees—or 7.25 percent. Of the total support contractor force of 249 minorities, contractors employed only 109 in technical fields.[36]

Evaluations of the impact of Marshall's EEO efforts consistently characterized them as a general failure. A critical evaluation of the impact of the space program on Huntsville by New York University graduate student Robert A. Myers found that the continually decreasing black population in the city had been "completely bypassed by Huntsville's rapid economic growth" and had "not participated equally with the white community in reaping the benefits" of the city's expanding economy.[37] As NASA worked to institute a more centralized EEO program from headquarters, it pointed out Marshall's failure to meet even "modest goals" and that the program suffered from an "extreme lack of proper staff and management support." NASA management continued to ignore the repeated efforts to bring Marshall's poor performance to its attention.[38] Discrimination inside the gates at Marshall was also a problem, as the director of NASA EEO, Ruth Bates Harris, remarked in May 1972 that she had "never seen a more unhappy, more disillusioned group of blacks."[39]

Over the decade of the 1960s, Marshall enacted several programs and worked with community organizations to increase the number of black workers at the center. The direction and scope of those programs changed from time to time, but many elements remained at the core. Recruiting qualified black applicants from the limited number of black engineering institutions across the country remained problematic, as black graduates earned more money working at private industrial firms located outside the overtly racist South. Alabama governor George Wallace's fervent stance against desegregation intensified what was an already difficult situation. At the same time, a revolution was unfolding beyond the gates of the Marshall Space Flight Center. The revolution included a civil rights battle in the city, an educational transformation at local HBCUs as administrators shifted from preparing students for limited professional opportunities to programs that embraced larger trends in the economy, and an attempt by local elites to assert hegemony over the channels of black participation in all areas.

Notes

1. Wilcock was one of five founders of the *Village Voice.*

2. Marshall Keith to Wilcock, Papers of the NAACP, Part 20: White Resistance and Reprisals, 1956–1965, Series: Group III, Series A, Administrative File: General Office File–Reprisals.

3. Roger Launius, *Project Apollo: A Retrospective Analysis* (Washington, DC: NASA, 1994), 7.

4. The "Space Crescent" was the common expression for the moon-shaped arc of NASA installations in the South that stretched from Cape Canaveral to Houston and included the Kennedy Space Center, Marshall, the Michoud Assembly Facility (New Orleans), the Mississippi Test Facility (Hancock County), and the Manned Spacecraft Center (Houston).

5. Executive Order 10925, March 6, 1961, https://www.eeoc.gov/eeoc/history/35th/thelaw/eo-10925.html.

6. NASA HQ Briefing Memorandum, April 1963, Digital Files, Marshall Space Flight Center (MSFC) Historical Reference Collection, Huntsville, AL.

7. "Star Salute–Paul Styles, Industry Relations Official, Negotiates Agreements," *Marshall Star,* May 15, 1963, Digital Files, MSFC Historical Reference Collection, Huntsville, AL.

8. NASA HQ Briefing Memorandum, April 1963, Digital Files, MSFC Historical Reference Collection, Huntsville, AL. Styles was later named the director of the MSFC Manpower Office and became the center's first Equal Employment Opportunity officer in October 1971.

9. Bobby M. Wilson, *American Johannesburg: Industrial and Racial Transformation in Birmingham* (Lanham, MD: Rowman and Littlefield, 2000), 226.

10. Webb to von Braun, April 19, 1963, Digital Files, MSFC Historical Reference Collection, Huntsville, AL.

11. Von Braun to Webb, July 15, 1963, Digital Files, MSFC Historical Reference Collection, Huntsville, AL. On June 11, 1963, Alabama governor George Wallace stood in front of the schoolhouse door of the University of Alabama to block the admission of the first black students, Vivian Malone, James Hood, and David McGlathery. McGlathery was a mathematician in the Research Projects Division of the Nuclear and Ion Physics Branch at MSFC and had applied to the Huntsville campus of the University of Alabama. He enrolled in classes on June 12, 1963, just one day after Wallace's action in Tuscaloosa.

12. Huntsville Contractors Equal Employment Opportunity Committee Report, July 5, 1963, MSFC Director's Files, Box 1, NARA, Atlanta.

13. Oral history interview with Clyde Foster, April 23, 1990, Digital Files, MSFC Historical Reference Collection, Huntsville, AL.

14. Memorandum, Hobart Taylor Jr. to Charles A. Horsky, July 29, 1963, "Coordinating Program for Equal Opportunity Housing for Federal Employees," Lee C. White Office Files, Lyndon B. Johnson Presidential Library, Austin, document provided via email by LBJ archivist Allen Fisher.

15. Ibid.

16. Marion Kent replaced Paul L. Styles as Marshall compliance officer effective from April 1963. In January 1964, Styles took over the newly created position of labor relations director for the entire NASA agency. In that position, he advised NASA management officials at headquarters and at field centers on labor-management relations with a focus on labor-contractor relations, which continually threatened to delay or stop work on NASA contracts during the Apollo program. He also served as the NASA representative with the Department of Labor, the NLRB, the President's Missile Sites Labor Commission, and the Federal Mediation and Conciliation Service (*Marshall Star,* January 29, 1964).

17. Quoted in Andrew J. Dunar and Stephen P. Waring, *Power to Explore: A History of Marshall Space Flight Center, 1960–1990* (Washington, DC: NASA, 1999), 120.

18. Kent to Gorman, November 22, 1963, Dr. Wernher von Braun Papers, Box 406, Folder 11, United States Space and Rocket Center (USSRC) Archives, Huntsville, AL.

19. Gorman to Hodgson, March 14, 1964, Digital Files, MSFC Historical Reference Collection, Huntsville, AL.

20. Memorandum by Lee White, June 12, 1964, Digital Files, MSFC Historical Reference Collection, Huntsville, AL.

21. Oral history interview with Charles Smoot, August 2, 2016.

22. Kent to Gorman, November 22, 1963, Box 406, Folder 11, Dr. Wernher von Braun Papers, United States Space and Rocket Center (USSRC) Archives, Huntsville, AL.

23. Memorandum, Macaluso to Taylor, May 9, 1964, Box 9, Hobart Taylor Jr. Papers, Bentley Historical Library, University of Michigan, Ann Arbor.

24. Rhett was a prominent Huntsville businessman from one of the older families, while Halsey was a wholesale food distributor.

25. Memorandum, Macaluso to Taylor, May 9, 1964, Box 9, Hobart Taylor Jr. Papers, Bentley Historical Library, University of Michigan, Ann Arbor.

26. Webb to Von Braun, July 13, 1964, Digital Files, MSFC Historical Reference Collection, Huntsville, AL.

27. Memorandum, Webb, July 13, 1964, James Webb Papers, NASA HQ Historical Reference Collection, Washington, DC.

28. James Webb to Lee White, July 13, 1964, Digital Files, MSFC Historical Reference Collection, Huntsville, AL.

29. Pencil notes, July 22, 1964, Box 3, David Newby Papers, University of Alabama Huntsville Archives, Huntsville.

30. Von Braun to Webb, August 11, 1964, Digital Files, MSFC Historical Reference Collection, Huntsville, AL.

31. Report from the Office of the White House Press Secretary, undated, Box 9, Hobart Taylor Jr. Papers, Bentley Historical Library, University of Michigan, Ann Arbor.

32. McMillan to Taylor, November 9, 1964, Box 9, Hobart Taylor Jr. Papers, Bentley Historical Library, University of Michigan, Ann Arbor.

33. Ibid.

34. Von Braun Speech to Alabama Chapter of the International Association of Public Employment Services, Huntsville, Alabama, May 13, 1966, Digital Files, MSFC Historical Reference Collection, Huntsville, AL.

35. Paul Styles to Dr. Eberhard Rees, May 5, 1971, MSFC Archives, EEO Collection.

36. "MSFC Contractor Black Employments," May 25, 1971, MSFC Archives, EEO Collection.

37. Robert A. Myers, "Planning for Impact: A Case Study on the Impact of the Space Program on Huntsville, Alabama" (Master's thesis, New York University, 1967).

38. Quoted in Dunar and Waring, *Power to Explore*, 124.

39. Ruth Bates Harris to Dr. Eberhard Rees, May 26, 1972, NASA MSFC Digital Collections.

PART III

INTERNATIONAL CONTEXT

Arnaldo Tamayo Méndez and Guion Bluford

The Last Cold War Race Battle

CATHLEEN LEWIS

There are two men who contend as the first person of African descent in space. Neither Arnaldo Tamayo Méndez nor Guion S. Bluford Jr. set out for this competition, nor was it a focal point for either's preparation for spaceflight, but the legacies of centuries of the transatlantic slave trade and the twentieth-century Cold War competition between the United States and the USSR placed them both in the competition. The legacy of the transatlantic African slave trade in the Western Hemisphere has been a shadow that lingers over North and South America. For a few years in the early 1980s, it hovered 200–300 miles above Earth. The Cold War–era superpower politics took the issue into space for reasons distantly related to race and slavery as a stark reminder that the shadow persists. In the end, this brief period revealed three things about race and the Cold War. First was the extent to which the uncomfortable legacy of African slavery remains prominent in the Western Hemisphere and American domestic politics. Second, this sequence of events in the 1980s also revealed that the rest of the world could be almost entirely oblivious to an issue that weighs so heavily on this side of the world. Also, finally, was the effectiveness with which the United States had dealt with the international scope of racial issues from the beginning of the civil rights movement in the South to the early 1980s while missing internally set metrics of accomplishment for integrating the workforce at NASA.

In reference books, exhibits, and public lectures, whenever mention occurs of the July 1983 flight of Guion (Guy) Bluford on board the US space shuttle *Challenger*, there is usually a small qualification, an asterisk, or a question or objection made. Bluford was the first US citizen of African descent to fly in space, but he was not the first black person, the qualification goes. That distinction belongs to Cuban pilot Arnaldo Tamayo Méndez, who flew in space on board a Soviet spacecraft launched from the Soviet Republic of Kazakhstan in 1980. There is no official arbiter or scorekeeper on this matter. The International Aeronautics Federation (IAF), the keeper of aviation and space world records, has no formal designation for the first of any given racial group flying in space or achieving any other aeronautical feats.[1] The participant-defined categories of records that the IAF keeps are measurable events according to geographical and political designations of astronautical feats, except for the categories of sex. The unofficial status of each man's race was set aside in the press, which felt the compelling need to mention race in the description of their flights. Race, a social and nonscientific construct, has no significance in human spaceflight, but it meant a great deal in the countries of origin of both men.

Beyond a common ancestry from Africa and the legacy of centuries of enslavement, Guion Bluford and Arnaldo Tamayo Méndez had little in common. Given the proximity of the United States and Cuba, it is surprising how different their lives had been. Cuba had freed its slaves two decades after the American Civil War, and that legacy had left an equally pernicious but more complex racial hierarchy in place in the island nation. Fidel Castro's decision to align the Cuban Revolution with the USSR applied an additional barrier between the two men, landing them on opposite sides of the Cold War. The result was two entirely different scenarios emerging as black astronaut candidates by the early 1980s.

The remarkable thing about Arnaldo Tamayo Méndez's role in this debate is that his race was not the burning issue at the time that he flew in orbit, at least not within the context of the Cold War. It was not until Bluford's flight three years later that Tamayo Méndez's identity within Cuba social strife became a subject of widespread discussion. Although not immediately apparent at the time, the elevation of Arnaldo Tamayo Méndez's flight in space to that of the first black person in space happened not at the time of his spaceflight, but after Bluford's. The story of this elevation is an exciting example of the interaction between domestic and international politics and its impact on the Space Race competition between the United States and the USSR. The story of how Tamayo Méndez became the first black man in space is an intricate story of how different forces of technology and national politics and social forces of two very different nations converged on two men. Moreover, one country's social issues leaped beyond the national boundaries and left a mark on an unrelated Cold War competition.

That distinction between citizenship and race carries weight due to the history of the United States. A generation earlier, blacks had been excluded from the flying opportunities that would have provided them with the credentials to stand as astronaut candidates for the Mercury program. Racial quotas had limited the number of black pilots during World War II, restricting their piloting activities to flying fighter aircraft and training for bomber units. As a result, by the end of the 1950s few blacks had qualified for test pilot school, which was a requirement for astronaut status. Only by the mid-1960s did the first generation of post–World War II, postsegregation black military pilots begin to expand. Guy Bluford himself had been a veteran of the Vietnam War, the first fully integrated war that the United States had fought. Within the United States, his symbolic role was evident.

Race was not the first social issue to emerge out of the US-Soviet competition in the Cold War/Space Race battlefield. The first instance of the interaction between domestic politics and the Cold War occurred two decades before the flights of the American space shuttle. In June 1963, Soviet-trained civil defense parachutist Valentina Tereshkova became the first woman in space. In response, there was an outcry among American women over the fact that the first woman in space was not an American, but a Russian, a Soviet citizen. The public debate was fueled by the less-public knowledge that women had indeed participated in the same screening that the Mercury astronauts had, and Tereshkova's flight ignited a long-running debate over gender roles in aviation and spaceflight. Even though Tereshkova's flight had been a stunt along the lines of previous stunts in Soviet spaceflight of repeating similar missions in order to claim a new record, it sparked deep concerns in the United States that the country had fallen behind the Soviets in women's issues as well as aerospace technology. As much as Soviet propagandists tried to make the connection, there was little to compare Tereshkova's experience and contribution to that of pre- and World War II female pilots. The staged all-female flights of the 1930s had garnered popular press to promote the new Soviet aviation industry and reinforced national devotion to civil defense training in anticipation of war.[2] Soviet women pilots had played both transport and combat roles during the war. Tereshkova's flight had been a one-off experience, and even though her corps of female cosmonaut candidates remained active for years after her flight, there was not another Soviet woman launched in space until the United States had announced the selection of women astronaut candidates almost two decades later. In 1963, Tereshkova precipitated an eruption of discussion in the United States over equal rights and woman, pitting American hero John Glenn against *Time* and *Life* magazine reporter Claire Booth Luce. The discussion quickly vaporized into the antagonistic atmosphere of the Cold War.

Race and the USSR

Despite its claims of egalitarianism and allegiance to recently decolonized nations of Africa, the USSR suffered its own patch of racial strife in the 1960s. Throughout the twentieth century, the USSR had always been cordial hosts to Americans, both black and white, who came with an implicit or explicit critique of their homeland. After World War II and the death of Stalin, Nikita Khrushchev reached out to Third World nations, most significantly to youth groups, culminating in the World Youth Festival held in Moscow in 1956.[3] From the Youth Festival grew invitations to almost exclusively male students from nonaligned countries to come to the USSR to study. The presence of invitees led to the formation of Friendship University, later known under the name of the assassinated Congolese prime minister Patrice Lumumba. All in all, there was a sevenfold increase in sub-Saharan African students studying in Moscow between 1959 and 1961. By the time of the collapse of the USSR, the enrollment numbers would reach a peak of 30,000.[4]

Close to the end of December 1963, there were public demonstrations in Moscow unlike others before. African students joined in protest over the death of a Ghanaian student, whose body was found in an isolated region of Moscow. After initial denials that a murder had taken place, accusations became known that the death of Edmund Assare-Addo took place during a violent dispute over mixed-race dating. The accusations and response were as every bit as grisly as one could expect in the Jim Crow United States, including Russian vigilante gangs that sought to re-solve disputes on their own terms.[5] Throughout the Cold War, accusa-tions of racism, cultural impropriety, and a general unease between the local Russian population and their guests would churn up political dis-turbances. However, unlike the few American blacks who traveled to the USSR during the 1920s and 1930s to escape Jim Crow, there was never a claim of an offer of assimilation to African students.

Race in Cuba

Given their similar experience in the African slave trade, one might ex-pect Cuba and the United States to have faced similar situations concern-ing race by the middle of the twentieth century. Enslaved Africans were brought to work the sugar plantations in Cuba as late as 1867 and were only emancipated in 1886 when African Cubans outnumbered European Cubans. The legacy of the American invasion and generations of upris-ings and revolutions left the slavery-constructed racial divide in Cuba as the status quo.

In postrevolutionary Cuba, Castro initially chose to ignore the issue of race until the poor treatment of black Cuban guerrillas came to his at-tention. However, his inaction did not counterbalance his awareness that black Cubans had not been the most enthusiastic participants in the revo-lution. He was not above making race into an international political state-ment, however. Castro's famous visit to the United Nations in 1960 and his decision to stay in Harlem was a powerful statement and established an iconic image of the transracial Cuban revolutionary. These images did not hide the fact that Castro himself was an opponent of the American Black Power movement in the 1960s and 1970s. Using race in international politics was one thing; supporting a movement that could have potentially explosive consequences at home was another.[6]

Cuba decided to intervene in the Angolan civil war in 1977, which res-urrected Castro's use of race in foreign relations. The ideological struggle in Angola drove Cuba's intervention, but the Cuban government did not evade the opportunity to demonstrate Cubans' racial link to the Angolans as a means of buttressing their support. They went as far as overrepresenting Afro-Cubans among the initial deployment of troops, although this might have been a method to disguise the presence of foreign troops in an African country.[7] However, as was the case of Cuban relations with black Americans, the opportunity to create an international stir was not equated with changes to the social system of race on the island.

US Civil Rights and the Space Program

Historians' exploration of the role that race has played in the Cold War is a relatively recent phenomenon. While equal rights for women has been a constant subtheme of twentieth-century modernization in Europe and the United States, race was a domestic issue that maintained a constant presence in the Western Hemisphere during the Cold War, especially for the United States.[8] The legal historian Mary Dudziak has written on the history of race in the Cold War.[9] Dudziak bases her historical assumptions about the relationship between race and the Cold War on Gunnar Myrdal's landmark history, *An American Dilemma*, basing the legal history on his assertion that the civil rights movement began in earnest during World War II.[10] According to Dudziak's extension of Myrdal's arguments, the American racial dilemma became a moral dilemma for American foreign policy during the world buildup to World War II. During the war years, it acquired tremendous international implications, driving the United States' limited forays into equal opportunity during the war and causing public discomfort in the vast areas that went unchanged. According to Dudziak, this American international self-consciousness about race drove the Roosevelt administration to respond more readily to domestic threats of protest than it had in the past. Roosevelt signed Executive Order 8802 with little enforcement strength to prohibit discrimination among US contractors to the War Department. The War Department tried experimental, segregated black units of the Army Air Forces and Marine Corps. The resulting lessons were mainly about the inefficiency of segregation and less about the rewards of integration, but the wartime actions established that the United States would have to take action if it was to maintain its postwar role of guardian of democracy.

By the beginning of the Cold War, Americans were committed to telling a particular story about civil rights—one that linked civil rights and progress by telling the story of a triumph of good over evil. By necessity, this civil rights story maintained US moral suasion after the war. Citing Michael Sherry, Dudziak points out that safeguarding the United States' overseas image was of prominent importance, even in the case of domestic policy. It was that motivation more than anything that contributed to the quick turnaround in the organization of American forces overseas and led to the deployment of integrated troops during the Korean War, just five short years after the segregated World War II.

Dudziak makes a three-pronged argument on the role that US civil rights played in the international arena that drove foreign policy. First was the importance to the American identity of the narrative of progress mentioned above. Second was concern over the experiences of brown-skinned foreigners visiting the United States, especially those coming from the newly independent former colonial states. Otherwise, isolated incidents threatened the US image abroad as a country of opportunity. Observers of foreign relations recognized the significance of the perception of the US abroad. The journalist Edward R. Murrow wrote to President Johnson in January 1964, saying, "The progress of the civil rights movement in this country is of preeminent interest overseas, particularly in Africa."[11] The third was the ruthlessness with which the government repressed dissent from blacks in America. The official US response to the internal critique of the civil rights situation in the country was disproportionate to the tone of those critiques, interpreting them as attacks on the Johnson administration.[12]

No matter how stable this political justification might have been in the 1950s and early 1960s, the three-pronged approach unraveled in the 1960s in response to other events that could not be controlled from either the White House or the Pentagon. Events of 1963, including both the bombing of the Sixteenth Street Baptist Church in Birmingham, Alabama (August 15), and the March on Washington (August 28) shattered the administration's narrative of progress. The further unraveling of the United States' ability to manage the storyline continued in 1964. Johnson's signing of the Civil Rights Act was quickly followed by the Gulf of Tonkin Resolution to expand the Vietnam War. Vietnam displaced racial progress as the foremost symbol of the United States, even in the eyes of White House insiders.[13] The increasing and prolonged white southern violence against the Freedom Rides and Martin Luther King Jr's receipt of the Nobel Prize for Peace in December 1964 internationalized the coverage of the civil rights struggle. The same was true of the increasing popularity of Malcolm X both at home and abroad and his break with the Nation of Islam for broader Muslim identity.

The toppling of the delicate balance of civil rights in America drew Soviet interest. There were many Soviet attempts to make a proper link between the urban riots of 1968 in the United States to American policy in Vietnam.[14] With the change in administrations and political parties in the White House, there was a change in tone and direction in domestic policy. "Just as Vietnam had eclipsed civil rights as a defining issue affecting US prestige abroad, law and order had eclipsed social justice as a politically popular response to racial conflict."[15] In the end, the "Cold War imperative for social change . . . did not survive the length of the Cold War itself."[16] However, the history of race in America remained an international story, with the United States pointing to progress from the past as an indication of progress and the USSR identifying that same past as an indication of political frailty.[17] Nixon's domestic and international civil rights policy ignored domestic race relations and continued through the Ford administration. The Nixon-Ford White House's willingness to deal with the apartheid South African government was the clear expression of their Cold War perception that fighting the Soviets and their surrogates trumped human rights.[18] Nixon and Secretary of State Henry Kissinger were not deliberately supporting racism, but their priorities were to support anything that challenged the USSR. That included American Black Panther complaints about rampant racism in Castro's Cuba.[19]

It was not until the Carter administration that human rights abroad regained some of the currency in domestic American politics that it once had immediately after World War II. Carter had won a close victory to the White House with the votes of southern blacks. As president, Carter had to deal with the more complicated aspects of the legacy of race relations. He made his claim on the role that race relations would play in foreign relations with his appointment of Andrew Young as ambassador to the United Nations. However well intentioned Carter's policies had been at the beginning of his term, they were marred by events that occurred under his watch that ultimately weakened his presidency irreparably. The

Iranian hostage crisis and the Soviet invasion of Afghanistan doomed all of Carter's foreign policy to failure and set the stage for a racially indifferent Reagan administration.[20]

The entire history of NASA and NASA-influenced efforts to increase diversity was not entirely grim. The instances of indirect attempts to improve the situation especially among southern blacks do offer a few nuggets of accomplishments. Two recently published histories of the pre- and early NASA periods present insights on what NASA could and could not influence in the social sphere of America. The first is the story of how Lyndon Johnson sought to impose the Great Society concepts onto the South through the NASA centers in the South. The second is the remarkable account of how West Virginia's and Virginia's segregated societies contributed to the formation of an isolated group of black women who calculated orbits and trajectories for NASA.

Richard Paul and Steven Moss dug deep and worked against time to uncover the results of efforts to impose Great Society standards to the societies surrounding NASA centers, focusing mainly on the attempts to integrate educational institutions in the Deep South.[21] Paul and Moss tracked down some of the surviving black southerners who were the original recruits to study at previously segregated southern engineering schools near NASA centers. The two also made comparisons of the impact of NASA's policies on the professional and technical labor forces in Texas, Alabama, Mississippi, and Florida. Margot Lee Shetterly approached the story of NASA-surrounded segregation from within by uncovering the stories of the black women mathematicians who worked for the segregated contractor to NASA's Langley Research Center.[22] Shetterly's story focuses on the life story of Katherine Johnson, who had the good fortune of being in the right place at a time when universities and contractors were willing to meet the letter of the law within the confines of the Jim Crow regime. What Paul and Moss and Shetterly uncovered were not so much the successes of NASA's efforts to integrate but the presence of blacks, even in the South, who were ready and willing to seize opportunities with NASA once any arose. The human spaceflight programs of the post-Apollo era unfolded with this history of race and the Cold War as one of many backgrounds. By the late 1970s and early 1980s, other, more severe issues dominated the US-Soviet global competition and race, and human spaceflight had lost its prominence. In contrast, the domestic history of the United States, the Soviet Union, and Cuba loomed more substantial and immediate in the stories of how human spaceflight was unfolding at that time, less inhibited by global competition and more influenced by local politics and recent history.[23]

Arnaldo Tamayo Méndez

Soyuz 38's flight occurred as part of a long-term change in the direction of human spaceflight within the Soviet Union, although next events sparked the relations between the United States and the USSR. Once the USSR had backed off its manned lunar program, the country continued to work out the internal competitions within its space industry to focus on long-duration spaceflight in near-Earth orbit. In order to sustain long-duration flights, the Soviets had to overcome a technical obstacle that prevented them from maintaining humans in space for over three months. The batteries that operated the systems on board the Soyuz at that time had a maximum flight life of sixty days. In order to exceed what had been the long-duration spaceflight record set by American astronauts William Pogue, Edward Gibson, and Gerald Carr on board the Skylab 4 mission in November 1973 of eighty-four days, the Russians would have to replace the Soyuz spacecraft docked to the Salyut 6 space station every sixty days. Of course, the Soviet technicians could quickly send a robotic spacecraft to the space station every two months, but their flight doctors recognized the psychological benefits of having visiting crews come to the station every ninety days. The truth was that long-duration spaceflight was often dull and isolating, and having visitors would contribute to the crews' health.

While the obvious solution would be to fly existing Soviet cosmonauts to the station to facilitate the craft swap-out, Soviet space planners turned to another alternative that would further distract attention from the shortcoming of the technology by announcing a new program. Interkosmos, the Council on International Cooperation on the Study and Uses of Space, had been founded in 1966 under the auspices of the Academy of Sciences of the USSR. The council began as a domestic Soviet government organization with representatives from the Ministries of Foreign Affairs, Communications, and Health and other scientific research organizations. The initial purpose of the council was to oversee cooperative space agreements with other countries, such as India, France, the United

States, Sweden, and Austria. In 1970, eight other socialist countries joined Interkosmos (including the USSR)—Bulgaria, Hungary, GDR, Cuba, Mongolia, Poland, Romania, and Czechoslovakia—to form the organization that became widely known during the 1970s as the Soviet alternative to NASA and the European Space Agency.

Salyut 6, the first station with dual docking ports to accommodate two Soyuz spacecraft, was ready for launch at Baikonur on September 29, 1977. The first mission to the station, Soyuz 25 (crew Kovalenok and Riumin), failed to dock on its initial approach and had to return to Earth.[24] The first successful docking mission with the Salyut 6 space station was accomplished by the Soyuz 26 crew of Georgi Grechko and Yuri Romanenko, who stayed on board the station for ninety-six days.[25] Grechko and Romanenko greeted two visiting missions to Salyut 6 during this time. The first was the all-Soviet crew of Soyuz 27—Soviets Dzhanibekov and Makarov.[26] The second visiting crew was that of Soyuz 28, the first Interkosmos mission that included a cosmonaut from the Czechoslovak Republic, Vladimír Remek, along with mission commander Aleksei Gubarev. In quick succession after the Czechoslovak mission, in June and August 1978 the Polish and German missions succeed with the Soyuz 30 and 31 craft, respectively. The Salyut 6 space station was the beginning of what developed into over two decades of long-duration spaceflight experimentation. By the 1980s, the duration of these flights was routinely exceeding the operational life of the Soyuz ferry spacecraft.

The timing of the introductions of this seemingly easy and inexpensive success of the international guest visitor program coincided with a dramatic shift in US-Soviet foreign relations. On Christmas Eve 1979, Soviet forces invaded Afghanistan, unleashing American objections to their human rights practices and denunciation of their imperial aims in Central Asia. While the USSR's actions strengthened President Carter's campaign on human rights that led to the American boycott of the 1980 Summer Olympic Games in Moscow, the Soviet response was further intransigence. In the realm of human spaceflight, the USSR immediately and quickly announced the expansion of its guest cosmonaut program to include the next group of Interkosmos nations of Bulgaria, Hungary, Romania, Vietnam, Cuba, and Mongolia. Primary cosmonaut and backups were selected from each country.

Moreover, although Soyuz spacecraft launches were still not officially announced in advance at that time, there was an official announcement of

Vietnamese fighter pilot turned cosmonaut Pham Tuan's flight in advance, with little specific detail.[27] The launch of Vietnamese fighter pilot Pham Tuan on board Soyuz 37 on July 23, 1980, was the launch of the sixth international crew from the USSR. The flight signaled that the program had expanded beyond the Warsaw Pact and set space observers to speculate which of the USSR's socialist allies would be next in spaceflight.

Arnaldo Tamayo Méndez's life story follows a social trajectory that is common to postrevolutionary Soviet and Cuban heroes. He was born on January 29, 1942, in Guantanamo. Biographies of him are unclear about the precise circumstances of his birth and early childhood, but he was an orphan from youth. The only certainty from his early biography is that he was mixed-race, showing evident African ancestry in his features. Later in life he claimed that he had to shine shoes as a child, as that was the only work available to him under the Batista regime.[28] In 1961, he graduated from the Eheristo Rebelde Technological Institute. He received flight training at the Eiskoi (now Komarov) Air Force School. He completed his flying studies at the Military Flight Training School in the Soviet Union. Returning to Cuba, Tamayo Méndez graduated from the Gomez Advanced Training School for Armaments. He joined the Cuban Communist Party in 1967 at age twenty-five, a few years too young, by Soviet standards, but probably an average age for an ambitious young air force officer.

On September 19, 1980, Cuban air force pilot Arnaldo Tamayo Méndez was launched from Baikonur on board the Soyuz 38 en route to the Salyut 6 space station. Experienced cosmonaut Yuri Romanenko commanded the mission.[29] The described mission of the weeklong stay on board the space station was to conduct experiments related to Earth observation and natural resources. As designed by the mission, Tamayo Méndez and Romanenko joined station crew Leonid Popov and Valerie Riumin on board Salyut 6 and returned to Earth on board Soyuz 37, which had carried the Vietnamese pilot to the station two months prior.[30] Popov and Riumin were within two weeks of breaking an endurance record (over 175 days, the final stay totaled 185 days). Tamayo Méndez was the seventh non-Soviet citizen to launch on a Soyuz. The Soyuz 38 crew performed over twenty experiments primarily described as Earth resources observation to aid Cuban agriculture. After the flight, Leonid Brezhnev awarded Tamayo Méndez the Order of Lenin and Hero of the Soviet Union award—the routine prizes that cosmonauts received. After his flight Tamayo Méndez continued his military career, achieving the rank of general in the Cuban air force as director of foreign affairs of the Ministry of the Revolutionary Armed Forces of Cuba. He has remained active in retired life, with sporadic appearances meeting foreign delegations, most recently with senior Chinese military officials in 1990.[31]

Soviet press coverage of Tamayo Méndez's flight spread over several sections of the print press. Not only did the flight achieve the usual front-page coverage accorded to TASS reporters, but it broke the barriers of the usual cast of reporters. Reporting during the mission spread beyond the front page and routine science reporting to the political and international pages. The standard report of the Soyuz 38 mission benefited from enhancement surrounding Raul Castro's visit to the USSR, which coincided with the mission to the space station. The first published report of the Soyuz 38 launch made it into print in *Pravda* the day after the successful launch. Beneath the prominent photos of Tamayo Méndez and his commander, Yuri Romanenko, were the standard format and brief captions of their careers. The anonymous report from TASS listed Tamayo Méndez's first-class status as a military pilot, his membership in the Communist Party of Cuba, and his selection as an astronaut candidate on March 1, 1978.[32] After two days in orbit, experienced Soviet space reporter Vladimir Gubarev began reporting in some detail on the specific activities of the international guest cosmonaut on board the Salyut 6 station. Gubarev first reported on Tamayo Méndez's assignment in this Interkosmos mission to conduct observational experiments and photography in assistance to Cuban agriculture with the well-wishes of Soviet leader Leonid Brezhnev.[33] Gubarev's article opened the way to reporting in subsequent pages of that issue in *Pravda*, first publishing the full texts of congratulatory telegrams from both Brezhnev and Fidel Castro. Both leaders applauded the space station mission as an accomplishment of national and socialist revolution, referring to Tamayo Méndez as the first Latin American cosmonaut.[34] A page later in the same edition, foreign correspondent A. Trepetov filed a report on the response in Havana to the mission. Trepetov reported on euphoric national pride. He quoted locals who frequently cited Yuri Gagarin's 1961 visit to Cuba after his first flight in space as the source of Soviet-Cuban fraternity that had risen to new heights with the Soyuz 38 launch.[35] The routine political and international reporting climaxed with Raul Castro's arrival in Moscow on September 22 and culminated with the award of Hero of the Soviet Union and the

Gold Star to Tamayo Méndez on the 27th, the day after he landed after his mission.[36]

Guion S. Bluford Jr.

At the same time that Soviet scientists were making the best of the limitations of their early Space Race hardware, the Americans were preparing to introduce a new type of spacecraft, the space shuttle orbiter, into operation. The deployment of the space shuttle mandated a new type of astronaut to populate the craft and fulfill the mission of exploiting space after the Apollo program. As a result, there was a dramatic change in the requirements for spaceflight. The 1959 requirements for NASA astronauts were restrictive to limit the potential candidate pool. Candidates had to be younger than forty years old, shorter than 5 feet 11 inches (180 centimeters), in excellent health, have earned a BA or equivalent in engineering, have the status as a qualified jet pilot, be a graduate of test pilot school, and have performed 1,500 hours or more of flying time.[37] These criteria effectively reduced the applicant pool to white male test pilots. Of the 500 military pilots who qualified under these criteria in 1959, seven were selected.[38] As the early human spaceflight program in the United States developed, the selection criteria changed. In 1962, three changes were made to the astronaut requirements. The age limit to qualify as an astronaut was lowered to thirty-five, the height limitation rose to 6 feet (182 centimeters), and civilians were permitted to apply.[39] These changes still excluded all but the youngest Korean War veterans and had little effect on the numbers of qualifying women or minorities. In 1964, in anticipation of an advanced Apollo Moon exploration program, applications were invited from educational training for positions as scientist-astronaut.[40] In the early 1970s, when the applications were open for shuttle astronauts, the requirements further relaxed combat experience requirements but increased the intensity of competition with more stringent academic requirements.

The necessary shuttle astronaut requirements from 1978 included excellent health, a BA in engineering or physical science or mathematics, 1,000 hours of flying time, and one of the following: an advanced degree or jet aircraft and flight-testing experience.[41] The more stringent academic requirements and more relaxed flight-training requirements had the effect of expanding the applicant pool outside of the white male bastion.

There was a significant pool of women and blacks who qualified and could apply.

Guion Bluford was among those who applied and was accepted into the shuttle astronaut program. Bluford was born in Philadelphia on November 22, 1942. He graduated from Pennsylvania State University in 1964 with a degree in aerospace engineering and was a distinguished air force Reserve Officers' Training Corps (ROTC) graduate. Bluford was a third-generation college graduate. His father had been a mechanical engineer.[42] Bluford's accomplishments were an impressive feat considering that the military services were segregated at the time of his birth. He continued his studies while in the air force after receiving his pilot's wings in 1966. He earned a doctor of philosophy in aerospace engineering with a minor in laser physics from the Air Force Institute of Technology in 1978, the year that he was accepted into the astronaut program.

The January announcement of the 1978 astronaut class, of which Sally Ride and Guy Bluford were members, proclaimed the strength in the diversity of minds as well as culture and gender. Six women were in the "Class of 1978." Three black men were among the thirty-five. "Diversity" was the catchword of this class, but gender and race were not the only indicators of diversity. As one journalist wrote, describing this new group of astronauts: "The diverse educational background of the astronauts also marks a departure from the nature of previous groups. This new diversity was due to the unique requirements of the new space shuttle program, for which the new astronauts have been training since July."[43]

At the time, the coincidence of an increasingly diverse astronaut corps and the spaceflights of cosmonauts from Asian and Latin American countries seemed to go unnoticed in the press. Accounts of Tamayo Méndez's flight noted the oft-repeated chorus of socialist goodwill among Soviet-aligned nations. Only one contemporary account of the flight of Soyuz 38 made a passing remark about Tamayo Méndez's ethnicity at the time of his flight. An Associated Press brief on September 18 noted that the Soviet-published photographs of Tamayo Méndez "suggest he was of mixed black and Caucasian origin."[44] All other references to Tamayo Méndez's race in the American were retrospective, after Bluford's flight. In contrast, Soviet references to Bluford's race were overt, citing him as the first black American included in a US spaceflight.[45]

How Did the Flights Come to Symbolize a Racial Standoff?

How did Tamayo Méndez's race become significant? It was only in the aftermath of discussing the significance of Sally Ride's flight to become the first American woman in space almost twenty years after Valentina Tereshkova's flight that attention turned to the scheduled flight of Guy Bluford, who was to be the first black among the "Class of 1978" to fly on board the *Challenger* in July 1983. The first journalist to link the flights of Tamayo Méndez and Guy Bluford was William Broad of the *New York Times*, who dispensed with finer, Latin American distinctions among blacks and mulattos and proclaimed Tamayo Méndez the first black in space as he fit the American classifications for racial identity.[46] Broad was quick to point out, however, that unlike Tamayo Méndez, Bluford had three other black American astronauts waiting to follow him into space.[47] There was a small irony that one of the first newspapers to challenge mainstream American assumptions about race and spaceflight was Bluford's hometown black newspaper, the *Pittsburgh Courier*, which stated outright that Tamayo Méndez was the first.[48]

One can assume that the *Pittsburgh Courier* was articulating a common sentiment about the role of race in spaceflight—if Valentina Tereshkova's flight had been a stunt to show up the Americans, then the flight by anyone who would be perceived, by American standards, to be black prior to Bluford's flight had to be a similar stunt. However, the Cuban population is as much as 60 percent of what Americans would perceive to be black; thus Tamayo Méndez's complexion is not so surprising. For Cubans, it would be more surprising if he had turned up as a manager of a foreign currency hotel in Cuba or well placed within the country's hospitality industry.[49]

So that is how, three years after his flight, Arnoldo Tamayo Méndez became the first black person in space. Surprisingly, that did not occur by birth on his part but happened mainly in the American press and as a result of American domestic politics and culture that were unrelated to the Cold War competition between the United States and the USSR. After three decades of promoting the idea of racial progress to bolster the American image abroad, especially in African nations and other developing nations, the United States had slowly abandoned its policy of promoting national progress as an indication of US superiority. Dissenting voices from the civil rights movement and the increasing involvement in Vietnam had made management of the message impossible. Nonetheless, the Cold War between the United States and the USSR continued. It is significant that during the Cold War, when the United States had not promoted its progress on civil rights, the Soviet Union had scored early points in women's issues—the Space Race. While the Johnson administration and NASA policies to improve the status and numbers of blacks among NASA engineering centers, especially in the Deep South, had made some changes, NASA had made no self-conscious effort to break through the gender or race barriers that surrounded the most public faces of the Space Race—the astronauts. While the Soviets dipped into their pool of civil defense–trained women for cosmonaut candidates, the Americans held fast to their initial description of an astronaut as a subset of experienced test pilots. The degree of successful accomplishment of these feats was less a measure of technological prowess and more a measure of the degree to which each side was willing to manipulate resources in the midst of international competition.

The biographies of Tamayo Méndez and Bluford, too, represent incongruent stories traveling on inconsistent social trajectories. Tamayo Méndez's story follows the predictable arc of impoverished origins overcoming racial and social constraints to high accomplishment, in no small part thanks to the Cuban Revolution. In contrast, Bluford's story is more challenging to fit into a single narrative neatly. On the one hand, as a black man in a previously all-white profession, he is a personification of social progress in America. On the other hand, as a third-generation college graduate, Bluford also represents an American ideal of hard work and perseverance, and the American Dream of improving status over generations. Tamayo Méndez and Bluford personified two different stories: one was held up as a rarity in support of a revolutionary idea, the other one was sold softly as a common modern-day occurrence. The more significant distinction was that Tamayo Méndez was a single exemplar while Bluford was one of three and then four, giving the higher strength of the American narrative through repetition. As an important side note, it must be remarked that the USSR, the twentieth-century homeland of Pushkin, did not try to find a descendant of an African in Russia to send into space, but relied on its only close ally with a large African diaspora to supply the cosmonaut.

Conclusion—Why Was This the Last Fight?

The propaganda currency of the spaceflights of Tamayo Méndez and Bluford differs from that of the early Space Race, especially from the flight of Valentina Tereshkova, in both planning and response. In the 1960s, it was clear that both sides shared a degree of focus on international prestige during the Cold War Space Race. Each side sought to demonstrate its social and technological sophistication through the flights. Two decades later, the sides were not competing at the same level for the same stakes. The USSR was using its Vietnamese and Cuban allies as a distraction from the 1979 Soviet invasion of Afghanistan. However clumsy the announcements of these flights had been, they were not a direct attack on the United States but were an attempt to distract the world from the Soviet foreign policy. The fact that they reminded the world of American struggles in both foreign and domestic policy was an added benefit, not a primary objective. On the American side, the response was profoundly different than it would have been twenty years earlier. Instead of the isolated but intense self-examination that resulted from Tereshkova's flight and the fierce repression of dissent that had characterized official policy during the early years of the civil rights movement, the announcement of the Cuban's flight elicited little official response from its neighbor to the north. Ambivalence over the significance of the pilot's racial identity was followed by indifference to his significance to the Cold War.

Tamayo Méndez's and Bluford's flights were the last of the Space Race's head-to-head competitions, because that was the last time that space competition could be a direct surrogate for each side to demonstrate its values for a onetime event. The missions betrayed vast gulfs in social conflict among these nations. Human spaceflights were no longer a matchup between the technologies of the United States and the USSR. The missions turned into a comparison between the social settings of each nation involved. The act of plucking a mixed-race pilot from among Cuban pilots seemed disingenuous on the Soviet Union's part. Previously, there had never been an Interkosmos guest cosmonaut candidate from a racial or ethnic minority.

Soviet and Cuban intentions withered even further when Tamayo Méndez's background was compared to that of Bluford. The American was not cherry-picked for his mission but was a member of a deep bench of racially and gender-diverse astronauts. Neither had he been cultivated through either the Lyndon Johnson–invoked political strategy of integrating the space program in the 1960s or the fruitless 1970s effort to address the ethnic composition of NASA contractors and staff. Bluford, along with his colleagues Frederick Gregory, Ronald McNair, and later Charles Bolden, had emerged into the astronaut program through broader systems of progress than NASA acting alone. Overall, three factors led to Bluford's selection as the "first" for the United States. The first factor was the more significant opportunity for blacks that the North had offered. Bluford and Gregory were descendants of the Great Migration, which brought their families north a generation before them. Even McNair and Bolden, who had each flourished in the isolated systems of South Carolina, sought educational opportunities in integrated northern schools. The second factor was the degree to which the military played the role of a social change agent in the 1960s. Three of the class of 1978—Bluford, Gregory, and Bolden—benefited from the integration of the armed forces, flying aircraft in America's first fully integrated war in Vietnam. Moreover, the final factor was the changed definition of an astronaut that occurred because of the shuttle program. This transformation from the heroic, Cold Warrior spacefarer to true explorers forced a pull away from military stereotypes and allowed a more diverse astronaut crew.

The long civil rights movement generated change for the astronaut program. Significantly, it brought to light the fact that while he might have seen the second black person to fly in space, Bluford represented profound changes in race relations in the United States over the course of a quarter of a century. Of course, the question remains as to whether this last Cold War race battle was inevitable or if somehow it was orchestrated. The lack of evidence that scheduling Tamayo Méndez's flight as an overt challenge to the freshly desegregated NASA astronaut corps is not proof that it was not the case. It is reasonable to assume that the USSR could have responded to the NASA announcement by requesting that the Cubans cherry-pick a mixed-race cosmonaut candidate for the Interkosmos mission. Although this scenario remains plausible, it does beg two additional questions. First, if they did so, why did the Soviets or the Cubans not guarantee the deal with a mixed-race backup for the mission?

Given that Soviets train and qualify crews as a unit, they increased the odds against themselves by having only one black man among the two crew options. The second and more pivotal question is why, if this had been a final play to one-up the Americans, did no one on either the

Soviet or Cuban side mention the fact at the time? The Soyuz 38 flight occurred during one of the steep declines of US-Soviet relations: within nine months of the Soviet invasion of Afghanistan and a few months of the US-led boycott of the Moscow-hosted Summer Olympics in 1980. If there was ever a time for the USSR to use propaganda ammunition against the United States, this had been the time.

Notes

1. The list of Interkosmos missions in order: Czechoslovakia (1978), Poland (1978), GDR (1978), Bulgaria (1979), Hungary (1980), Vietnam (1981), Cuba (1980), Mongolia (1981), Romania (1981), France (1982), India (1984), Syria (1987), Bulgaria (1988), and Afghanistan (1988). The list of black Americans who have flown in space: Bluford (1983), McNair (1984), Gregory (1985), Bolden (1986), Harris (1991), Jemison (1992), Winston Scott (1996), Robert Curbeam (1997), Michael Anderson (1998), Stephanie Wilson (2006), Joan Higginbotham (2006), B. Alvin Dew (2007), Leland D. Melvin (2008), and Robert Satcher (2009).

2. Lazar Brontman. *The Heroic Flight of the Rodina* (Moscow: Foreign Languages Publishing House, 1938).

3. Maxim Matusevich, "Testing the Limits of Soviet Internationalism: African Students in the Soviet Union," in *Race, Ethnicity, and the Cold War: A Global Perspective*, ed. Philip E. Muehlenbeck (Nashville, TN: Vanderbilt University Press, 2012), 145–165.

4. Ibid.

5. Julie Hessler, "Death of an African Student in Moscow: Race, Politics and the Cold War," Cahiers du Monde Russe, vol. 4, nos. 1/2, *Repenser le Degal: Versions du Socialisms influences internationals et société sovietique* (January–June 2006): 33–63.

6. Henley Adams, "Race and the Cuban Revolution: The Impact of Cuba's Intervention in Angola," in Muehlenbeck, *Race, Ethnicity, and the Cold War*, 206.

7. Ibid., 210.

8. Susan Bridger, "The Cold War and the Cosmos: Valentina Tereshkova and the First Woman's Space Flight," in *Women in the Khrushchev Era*, ed. Melanie Ilic, Susan E. Reid, and Lynne Attwood (New York: Palgrave Macmillan, 2004), 222–237.

9. Mary. L. Dudziak, *Cold War Civil Rights: Race and the Image of American Democracy* (Princeton, NJ: Princeton University Press, 2000).

10. Gunnar Myrdal and Sissela Bok, *An American Dilemma: The Negro Problem and Modern Democracy* (New Brunswick, NJ: Transaction Publishers, 1996).

11. Dudziak, *Cold War Civil Rights*, 209

12. Ibid., 210–220.

13. Ibid., 241.

14. Ibid., 243.

15. Ibid., 248.

16. Ibid., 249

17. Ibid., 253–254.

18. Thomas Bortstelmann, *The Cold War and the Color Line: American Race Relations in the Global Arena* (Cambridge, MA: Harvard University Press, 2001), 239–241.

19. Ibid., 242.

20. Ibid., 252–259.

21. Richard Paul and Steven Moss, *We Could Not Fail: The First African Americans in the Space Program* (Austin: University of Texas Press, 2015).

22. Margot Lee Shetterly, *Hidden Figures: The American Dream and the Untold Story of Black Women Mathematicians Who Helped Win the Space Race* (New York: William Morrow, 2016).

23. Kim McQuaid, "Race, Gender, and Space Exploration: A Chapter in the Social History of the Space Age," *Journal of American Studies* 41, no. 2 (2007): 405–434.

24. Iurii Pavlovich Semenov, ed., *Raketno-Kosmicheskaia Korporatsiia "Energiia" Imeni S. P. Koroleva* (Korolev: Raketno-kosmicheskaia korporatsiia "Energiia" imeni S. P. Koroleva, 1996), 298.

25. December 10, 1977, to March 16, 1978.

26. March 2–10, 1978.

27. The Soviets released information about the flight of a Vietnamese cosmonaut before releasing an official press release. The announcement occurred during the 1980 Olympics in Moscow, at a time when many journalists were in the country covering the athletes. David K. Willis, "Soviets Bid for Third World Prestige with Viet Cosmonaut," *Christian Science Monitor*, July 25, 1980.

28. Jack Sheehan, "Across the U.S.-Cuba Divide, A Retired General Takes a Step," *Washington Post* (1974–Current File), May 3, 1998, 1, http://search.proquest.com/docvie w/1620165112?accountid=46638.

29. "Cuban in Crew of Soyuz Craft Sent into Space," *Washington Post* (1974–Current File), September 19, 1980, 1, http://search.proquest.com/docview/147186338?accou ntid=46638.

30. "Bread and Salt Greet 2 Visitors to Space Lab," *New York Times* (1923–Current File), September 21, 1980, 11, http://search.proquest.com/docview/121093025?accountid =46638.

31. Sheehan, "Across the U.S.-Cuba Divide," 1.

32. "Kosmonavt-issledovatel' korablia 'Soiuz 38' podpalkovnik Arnal'do Tamaio Mendes" (Research cosmonaut of Soyuz 38, Lt. Col. Arnaldo Tamayo Méndez), *Pravda* no. 263 (22693), 1.

33. V. Gubarev, "Uzhin na reassvete" (Dinner at Dawn), *Pravda* no. 265 (22695), 3.

34. "Primer druzhba i sotrunichestva" (Example of Friendship and Cooperation), *Pravda* no. 265 (22695), 4.

35. A. Trepetov, "Kuba likyet" (Cuba Rejoices), *Pravda* no. 265 (22695), 5.

36. "Priem L. I. Brezhnevym R. Kastro" (Reception of Raul Castro by Leonid I. Brezhnev), *Pravda* no. 267 (22697), September 23, 1980, 1; "Ukaz" (Decree), *Pravda* no. 271 (22701), September 27, 1980.

37. "Astronaut Selection and Training," *NASA Facts* (KSC 35-81, revised May 1985), 2.

38. The original seven American astronauts were: Alan Shepherd, Gus Grissom, John Glenn, Scott Carpenter, Wally Schirra, Gordon Cooper, and Deke Slayton. Slayton did not fly a Mercury mission due to a heart murmur but later flew on board the Apollo spacecraft in the Apollo-Soyuz Test Project mission in July 1975.

39. "Astronaut Selection and Training," 2.

40. Ibid.

41. Ibid.

42. Penelope McMillan, "Black Shuttle Astronaut Sees Self as 'Role Model,'" *Los Angeles Times* (1923–Current File), July 14, 1983, 2, http://search.proquest.com/docview/153527665?accountid=46638.

43. Thomas O'Toole, "Shuttle Poised for Milestone Mission," *Washington Post* (1974–Current File), August 29, 1983, 2.

44. "Cuban in Crew of Soyuz Craft Sent into Space," *Washington Post* (1974–Current File), September 19, 1980, https://search.proquest.com/docview/147186338?accountid=46638.

45. "Novyi polet 'Chellengzhera'" (New flight of "Challenger") *Pravda* no. 243 (23769), August 31, 1983, 5.

46. William J. Broad, "First U.S. Black in Space: Guion Stewart Bluford, Jr.," *New York Times*, August 31, 1983.

47. McNair, Gregory, and Bolden did follow Bluford into space each subsequent year through 1986 (Bolden's flight).

48. Robert Fikes, "First Black Astronaut," *New Pittsburgh Courier* (1981–2002), February 22, 1986, 1, http://search.proquest.com/docview/201772570?accountid=46638.

49. Sheehan, "Across the U.S.-Cuba Divide," 1.

The Congressional Black Caucus and the Closure of NASA's Satellite Tracking Station at Hartebeesthoek, South Africa

KEITH SNEDEGAR

When Charles C. Diggs Jr., one of the founders of the Congressional Black Caucus (CBC), visited South Africa in 1971, he paid a call at the National Aeronautics and Space Administration (NASA) satellite tracking station at Hartebeesthoek, northwest of Johannesburg. There he discovered a racially segregated facility where technical jobs were reserved for white employees; black Africans performed menial labor, had segregated toilets, and ate their meals under a tree. Upon his return to the United States, the Detroit congressman embarked on a two-year struggle, first for an explanation of NASA policies with respect to Hartebeesthoek, and later for the closure of facility. NASA administration under James Fletcher exhibited resistance to demands for change at the tracking station. Only after Representative Charles Rangel, another CBC member, proposed a reduction in NASA appropriations did the agency announce plans to end its partnership with South Africa in operating the tracking station. NASA's public statements suggested that a scientific rationale lay behind the station's eventual closure in 1975, but this episode clearly indicates that NASA was acting chiefly under political pressure from Diggs and

Rangel, while its management remained largely insensitive to global issues of racial equality.

The story of the South African tracking station is widely known and has been particularly well explored by Sunny Tsiao and Rebecca Eisenberg.[1] Tsiao's definitive institutional history of NASA's tracking and data networks gives wholesale coverage of the Hartebeesthoek station, while Eisenberg's Portland State University thesis places the station in the context of Cold War relations between South Africa and the United States. This essay attempts to highlight Hartebeesthoek in the nexus of civil rights and African liberation struggles. In addition, the essay reflects on the social conservatism of NASA administrators born perhaps out of a technocratic focus on the objectives before them.

South Africa, the Space Age, and Apartheid

From the very beginning of the Space Age white South Africans had an abiding interest in astronautics and related fields.[2] As part of the International Geophysical Year (IGY) (1957–1958), the Smithsonian Astrophysical Observatory organized Project Moonwatch to visually track the orbits of artificial satellites around the globe. Such was the local enthusiasm for the project that in 1958 the number of satellite observations contributed by South African participants was second only to those of Americans. South African scientific institutions actively sought out partnerships with American organizations engaged in space science, including NASA. The Republic Observatory, South Africa's national facility for astronomical research, participated in NASA's International Planetary Patrol Program in the 1960s. With a special camera on loan from the Lowell Observatory, astronomers in Johannesburg took photographs of Mars at hourly intervals to monitor the Red Planet in support of Mariner flyby missions. Under the auspices of Harvard University, and later the Smithsonian Institution, the Boyden Observatory near Bloemfontein was actively engaged in satellite tracking; its headline success came in photographically locating the lost communications satellite *Syncom I* in February 1962.

Nelson Mandela had slipped out of the country a few weeks before *Syncom I* was launched. His mission was to receive military training abroad and help organize *Umknonto we Sizwe,* "The Spear of the Nation," as an armed resistance to the white-minority government of South Africa. In the early 1960s, approximately three million South Africans of European descent enjoyed civil and political rights under a Westminster-style parliamentary government, while roughly thirteen million of their fellow African, Asian, and mixed-race countrypeople were denied basic rights and social opportunities in a legal system of segregation known as apartheid. The discrimination and degradation suffered by South African blacks was, if anything, more severe than that experienced by African Americans in the Jim Crow South.[3]

Deep Space Station-51 at Hartebeesthoek

In January 1961, two months before South Africa left the British Commonwealth of Nations in defense of its apartheid policies, the governments of the United States and South Africa entered into an agreement concerning the establishment of the Deep Space Instrumentation Facility (DSIF) at Hartebeesthoek for tracking and communicating with spacecraft on planetary missions. Funding and technical support would be provided by NASA, while daily operations were to be the responsibility of the South African Council for Scientific and Industrial Research (CSIR). The Hartebeesthoek station was one of thirty facilities spread across the globe, forming NASA's Spaceflight Tracking and Data Network (STDN), but was specially linked to the Deep Space stations at Goldstone, California, and Woomera, Australia. Located in a 4,000-acre plot of undeveloped highveld, Hartebeesthoek comprised three elements by 1964: an air force tracking facility, a Smithsonian-run Minitrack facility, and a NASA Deep Space Network facility, designated DSS-51. DSS-51 demonstrated its value on the Surveyor I mission when it was the first Deep Space Network station to acquire the departing vehicle after launch. "It was desirable to inspect the telemetry to verify the spacecraft's state of health as soon as possible," David Harland observed, "as the transmitter could not operate at high power for more than an hour without overheating."[4] Hartebeesthoek was also the Deep Space Network station that received the first close-up images of the surface of Mars taken by Mariner IV in July 1965.

At its height in the late 1960s, Hartebeesthoek employed 280 South Africans; none of the technical personnel were black.[5] Meanwhile, the sensitivity of the tracking station's political situation grew as the American civil rights movement gained momentum with the 1963 March on Washington and the Civil Rights and Voting Rights Acts that followed. This apparently was not lost on NASA officials, but the geographic location of Hartebeesthoek made it a desideratum, as the South African tracking station filled a longitudinal gap between the Americas and Australia. A national security report in 1964 suggested that the air force and NASA facilities at Hartebeesthoek not be publicized in order to prevent any negative backlash from the American public.[6] Activists in the American Committee on Africa were already well aware of the racial politics surrounding Hartebeesthoek. A July 1965 American Committee newsletter noted a speech by South African prime minister Hendrik Verwoerd in which he asserted that his country would enforce a whites-only policy with respect to employment at the tracking station. For the American Committee, this meant that NASA's use of Hartebeesthoek entailed complicity in apartheid racial discrimination. Congressman Charles Diggs was one of the influential people the American Committee made contact with concerning this issue.[7]

Charles C. Diggs Jr.

The son of a successful Detroit mortician and entrepreneur, Charles C. Diggs Jr. was the first African American sent to Congress from Michigan.[8] He was elected to the House of Representatives in 1954. Over a congressional career spanning twenty-five years, Diggs looked for ways to promote an African American legislative agenda. Frustrated by the lack of organization among African American representatives, Diggs founded the Democratic Select Committee (DSC) in 1969 to coordinate the efforts of black members. Newly elected black representatives in the 91st Congress (1969–1971), including Shirley Chisholm, William Clay Sr., and Louis Stokes, embraced the idea of building clout in the House and were amenable to joining an African American voting bloc. With Diggs as its first chairperson, the DSC was rebranded as the CBC in 1971. He proclaimed the CBC "the first departure from the individualistic policies that characterized black congressmen in the past." In addition to promoting a civil rights agenda at home, Diggs wanted the CBC to influence US policy toward Africa. Nicknamed "Mr. Africa" because of his commitment to African decolonization and to the advancement of newly independent states, Diggs participated in numerous official missions to the continent. At some personal risk to himself, in February 1969 he headed a fact-finding mission to Nigeria to investigate relief programs and to explore a possible ceasefire during the Biafran War. That same year Diggs secured his appointment as chair of the House Foreign Affairs Subcommittee on Africa in 1969. Through this chairmanship, which he held until 1978, he emphasized the importance of Africa in American foreign relations and sought to increase African aid programs.

With respect to South Africa, Diggs and his DSC/CBC colleagues in the 91st Congress introduced twelve bills calling for sanctions against the country's white-minority regime. None of these bills reached the House floor for a vote, but CBC members submitted another twenty-six sanction proposals in the 92nd Congress.[9] Diggs meanwhile orchestrated a series of subcommittee hearings to investigate how American corporations helped sustain the South African economy, despite official US antipathy toward the country's racist policies. The Michigan representative also organized a congressional delegation to South Africa to personally observe American commercial and diplomatic complicity in a system of racial oppression. His efforts to highlight US entanglements with apartheid led the South African government to bar Diggs from entering the country on later fact-finding trips. Diggs's initial objectives were to put an end to the importation of South African sugar and to pressure US corporations to divest from South African operations. However, with the exception of the CBC, political support for these moves was nonexistent. His one noteworthy success concerning South Africa in this period involved the closure of DSS-51 at Hartebeesthoek.

After his 1971 visit to South Africa, Congressman Diggs used his position as chair of the House Foreign Affairs Subcommittee on Africa to request information from NASA on the operation of DSS-51 and to call NASA administrators to testify before the subcommittee. The hearings themselves were chiefly arranged to probe the activities of American businesses in the apartheid state, but Diggs widened the purview of the subcommittee's work. Diggs believed that the economic relationships that helped sustain apartheid could be undermined, although many businesses operating in South Africa did not think the US government was serious in its objections to the racist state "because of the examples set by our own embassy—where they know, as a matter of policy, black Americans are not assigned—and by the NASA tracking station, where discrimination is practiced in its most blatant form."[10]

Associate Deputy Director Willis H. Shapley (son of the astronomer Harlow Shapley) and Gerald Truszynski, associate administrator for tracking and data acquisition, appeared before Diggs's subcommittee. Shapley testified that NASA had selected optimum or at least acceptable locations for tracking stations based on technical considerations. Hartebeesthoek met the necessary geographical criteria.[11] Diggs asked how South Africa would respond if NASA insisted on desegregated working conditions. Shapley stated that South Africans would likely close the station rather than allow any changes to the racialized environment. Further, relocation of DSS-51 would cost an unacceptable $35 million; it was operationally and economically unsound to realign the Deep Space Network, Shapley testified. Hartebeesthoek had worked "within the existing system" of apartheid and would, by implication, continue to do so. It is telling that Shapley offered no alternative to the South African link in the Deep Space Network. In the hearings and related correspondence with Congressman Diggs he omitted to mention DSS-61 at Robledo de Chevala, Spain, a primary Northern Hemisphere tracking station that could be used as a back-up to or a replacement for Hartebeesthoek operations. In his testimony to Diggs's subcommittee, Truszynski also underscored "the two basic facts concerning the NASA presence in South Africa," namely that Hartebeesthoek was in a geographically advantageous position and that NASA was not the direct employer of those who worked at the station.[12] Neither Shapley nor Truszynski gave much ground to Diggs's inquisition. At most, Shapley suggested that NASA would consider sending an interracial technical team to Hartebeesthoek on a routine visit.[13] This did not impress Diggs, who openly complained that it was intolerable for the administrative whims of South African government officials to determine the employment practices of a facility paid for by the American taxpayer.[14]

Publicity surrounding congressional criticism of American relations with the apartheid regime was becoming an impediment to NASA's business-as-usual attitude.[15] The agency's presence at Hartebeesthoek came increasingly under scrutiny. Larger tracking dishes and more sophisticated equipment being added to the Madrid Deep Space Communications Complex (MDSCC), of which DSS-61 was a part, undercut the leverage of Hartebeesthoek's strategic geographical advantage. The closure of the Hartebeesthoek Tracking Station was finally announced in July 1973.

Ironically, Charles Diggs's visit to Hartebeesthoek may have had a more near-term personal impact on some of the South African technicians than on NASA administrators. Colin Garvie recalls having been proud to work at the tracking station when he was first employed there in 1965. He felt that South Africa was making an important contribution to space exploration; for him, it was "the chance of a lifetime" to travel to the Jet Propulsion Laboratory and Goldstone for training in connection with the Mariner Mars missions. Little did he think of the injustice of the apartheid system or of its absolute exclusion of "many equally deserving South Africans" from the opportunities he enjoyed. Garvie's awaking was slow in coming. He acknowledged the reaction of the Anglican dean of Pretoria, the Reverend Mark Nye, to the Apollo program in 1969; landing on the Moon was an impressive achievement, but, Nye asserted, there should be a space program of the human spirit "to tackle the far more difficult problems of man in his relationship to himself, to society and to God." The turning point for Garvie came in August 1971, when Congressman Diggs made his inspection of Hartebeeshoek. That Diggs was a devout Christian (a Baptist) impressed Garvie, who was beginning to have doubts about the ethical standing of apartheid, and sensed a call to religious purpose. Diggs's visit, "unpopular as it was, precipitated my decision to move on. I had come to the end of my contractual commitments and obligations to Hartebeesthoek. I resigned and candidated for the Methodist ministry at the end of 1971."[16] Garvie's recollections underscore the fact that those who worked at the tracking station did not exist in a social vacuum; they were fully aware of the racialized environment in which they operated, even if most of them did not take any action to transform it.

The Role of Charles Rangel

A veteran of the Korean War, Charles ("Charlie") Bernard Rangel was first elected to Congress from New York's Eighteenth Congressional District in 1970. He held that office, several reapportionments notwithstanding, until January 2017. An early member of the CBC, Rangel served on the House Foreign Affairs Committee as well as the Astronautics and Space Technology Subcommittee. He soon came to an understanding with Diggs that political traction might be gained by pressing NASA to explain its South African connection.

Patrick Swygert, an aide to Rangel, reported that the congressman saw Hartebeesthoek as an important target for anti-apartheid activism. The tracking station was valuable to the apartheid regime, as it provided South African scientists an interface with international colleagues and access to American technology. Additionally, it was a revenue-generating unit of the CSIR, employing nearly 300 staff by 1972.[17] For its part, NASA was symbolic of American achievements in technology and human endeavor. Unlike, say, sugar or chrome importation from South Africa, the space program had a major public profile (although the tracking station in Hartebeesthoek station was largely unknown to the American public). That made criticism of NASA potentially impactful, possibly creating more leverage against the white-minority regime in Pretoria. With regard to the South African tracking station, Congressman Rangel repeatedly requested information from NASA on the agency's compliance with executive orders on nondiscrimination. Swygert argued that a response was forthcoming in 1971. "In 1972 after constant inquiries, again and again and again, saying when are you going to tell us whether you are in compliance with the legal requirements of the order, NASA sent us a very cleverly-worded legal document which attempted to say that yes, they were."[18] Rangel's office found an almost impenetrable bureaucracy. Between NASA and the State Department was a rabbit hole of excuses. There was little NASA could do about Hartebeesthoek because it was managed by South African officials, yet the South African government was not, Swygert was told, directly involved with the tracking station, as it came under the aegis of a separate entity, the CSIR. State Department officials did not readily admit that NASA had any means of influencing CSIR operations. In fact, shortly after Diggs initiated his hearings and Rangel his correspondence, and almost certainly in response to these actions, Deputy Assistant Secretary of State Robert S. Smith met with Shapley and Truszynski to discuss the need for improvements for black employees at Hartebeesthoek. Two weeks later Shapley and Truszynski flew to South Africa to consult with Frank Hewitt, vice president of the CSIR.[19]

In the words of Swygert, Rangel decided "to go for the jugular" and on the floor of the House of Representatives proposed an amendment to NASA appropriations cutting off funds for the tracking station. Rangel observed that Hartebeesthoek was "the only U.S. government-subsidized installation in the world where racial segregation prevails under law, from top management to the toilets."[20] Rangel's amendment was easily voted down, perhaps as much out of ignorance as anything else. Speaking against the amendment, the chair of the Science and Astronautics Committee, Olin Teague, argued that without the South African communication link, astronauts ran the risk of being lost in space; he seemed to be unaware that Hartebeesthoek did not regularly support manned space missions. Congressman Robert Leggett was more explicit in his opposition, stating, "I do not believe we should use the space program as a vehicle to vindicate civil rights theories or to use civil rights as a basis to kill the space program."[21]

Swygert thought it was miraculous that just prior to the 1972 appropriation vote, NASA and the CSIR announced that they had reached an understanding on a cafeteria, school, and recreation area for black African employees at the tracking station. In May 1973, Rangel again moved his amendment, this time garnering 104 votes in favor, while Edward Kennedy introduced a companion proposal in the Senate. Kennedy withdrew his amendment after Frank Moss, the chairman of the Senate space committee, made a commitment to hold further hearings into the operation of Hartebeesthoek.

NASA Administrator James Fletcher

James C. Fletcher became NASA administrator in March 1971. The Apollo program would conclude in the next year and a half, and, in an environment of declining budgets, Fletcher would dedicate himself to advancing the space shuttle as the next generation of manned spacecraft. The Hartebeesthoek tracking station was a relatively minor issue on his managerial agenda, but the potential for embarrassment it posed was not something Fletcher wanted to get out of hand. Through protracted correspondence, Congressman Diggs pushed Fletcher to explain NASA's position with respect to the Hartebeesthoek station.

Fletcher noted in his replies to Diggs that local improvement programs were having an impact on the quality of life for black employees, with the construction of (albeit segregated) staff housing, an elementary school, and a medical clinic. Diggs and Rangel, however, considered the changes mere window dressing. The apartheid work environment remained essentially unchanged. "Local improvements" had no impact on the gross inequality of the work environment at Hartebeesthoek. Fletcher politely replied to additional requests for information and clarification but offered more detailed responses through Shapley and Truszynski.

The conflict over DSS-51 should be seen in the context of struggles over civil and employment rights closer to home. Fletcher had shown himself to be resistant to equal opportunity hiring practices in his tactless handling of Ruth Bates Harris in 1973. Harris, the first African American woman to head NASA's Equal Employment Opportunity (EEO) office, had coauthored a report arguing that the agency's hiring equity practices were "a sham" and that NASA leadership was indifferent if not openly hostile to affirmative action. While Fletcher admitted that NASA's employment of women and minorities was "not a record in which we can take pride," he straightaway fired Harris and disciplined the other authors of the negative report.[22] In connection with Fletcher's personal response to affirmative action, Brenda Plummer has argued that his religious affiliation was a limiting factor. "His Mormon faith posed a problem for him, as the laws he was obligated to uphold conflicted with some of the tenets of his religion as the Church of Latter Day Saints (LDS) then interpreted them. The refusal at the time to admit blacks to the Mormon priesthood and views that clashed with feminist thinking posed a dilemma for Fletcher."[23] While Fletcher's membership in the LDS Church may have inclined him to certain opinions, there is no indication in his professional or personal correspondence that would support the view that religion shaped his attitudes toward civil or employment rights. His public record would indicate the contrary. As president of the University of Utah in the 1960s Fletcher successfully negotiated heated situations among faculty and students involving race relations. Also, when, as NASA administrator, Fletcher was the target of insinuations that he was obliged to follow the guidance of the Mormon prophet, he emphatically declared his adherence to the principle of separation of church and state.[24]

Writing in *Science* shortly after the firing of Ruth Bates Harris, Constance Holden observed that "the overwhelmingly white male domination of NASA is making it an increasingly conspicuous and embarrassing anomaly among government agencies. . . . It would appear that considerably stronger pressure than has hitherto been exerted is needed to get NASA to move."[25] Fletcher's actions basically mirrored the social assumptions of the engineering, scientific, and managerial culture in which they were embedded. For Fletcher, the space program was the realization of can-do optimism and the belief in technological progress founded on what his predecessor, Thomas Paine, had called "the triumph of the squares." Social criticism from Harris internally, and from Diggs and Rangel externally, introduced incoherence to an otherwise elegant, if technocratically narrow, vision. As Jennifer Manuse has commented in her study of the evolution of space communication networks, legacy constrains the operation of complex systems.[26] The legacy of white-male administration of white-male technical work was not something easily disrupted.

Whether or not the writing was on the wall for Hartebeesthoek's future, Fletcher, an experienced administrator, opted to cut NASA's losses. Practical considerations of whether to repatriate technical equipment or donate it to the South African government had to be dealt with. (The CSIR eventually acquired all of NASA's installations in the country.) Also, the continued operation of the station was desired in support of the Pioneer 10 flyby of Jupiter as well as the near-Earth segment of the Viking missions to Mars. Dated July 10, 1973, the NASA press release announcing the phased closure of DSS-51 emphasized the declining value of Hartebeesthoek in spacecraft communication. By the second half of the decade, it would not have a mission to fulfill. Confidential State Department instructions to the US embassy in Pretoria highlighted the political framing of the argument. "Thrust of all exchanges on subject will be to emphasize that NASA action re Johannesburg station dictated by technical program requirements only and to avoid any implication that decision result of other considerations."[27] Both Diggs and Rangel welcomed NASA's announcement, but they vocally disagreed with its suggestion that technical considerations had been paramount in the closure decision. Diggs issued a statement that NASA "responded to pressures by myself, Congressman Rangel of New York, and the Congressional Black Caucus, and our friends in the Senate, particularly Senator Edward Kennedy." Kennedy supported this contention. To him, the closure of the station implied that "there can be meaningful responses to those conditions that [are] in conflict with the basic principles that we stand for in this country."[28]

On the same day as the NASA press release on DSS-51, the British government announced its intention to resume the sale of military weapons to South Africa. Clearly, the victory of Diggs and Rangel had been a small one, and the path to meaningful sanctions against the apartheid regime would be long. In terms of the American civil rights movement, the Hartebeesthoek affair also paled in comparison with the Ruth Bates Harris controversy that came into public view at the end of 1973. Nonetheless, the efforts of CBC chairman Charles Diggs and his colleague Charles Rangel had demonstrated that institutional change was not to come to NASA without concerted external pressure. The heritage of white-male administration and white-male technical work meant that the emergence of a more inclusive vision of who might participate in the international endeavor to explore outer space was slow in coming.

Notes

1. Sunny Tsiao, *"Read You Loud and Clear!" The Story of NASA's Spaceflight Tracking and Data Network* (Washington, DC: NASA History Division, 2008); Rebecca Nicole Eisenberg, "Reexamining the Global Cold War in South Africa: Port Usage, Space Tracking and Weapons Sales" (MA thesis, Portland State University, 2012).

2. P. Martinez, "Space Science and Technology in South Africa: An Overview," *African Skies* 12 (2008): 46. A few black students studied physics at the segregated University of Fort Hare, where there was an officially sanctioned astronomy club in the 1960s. But to the best of my knowledge, the response of black South Africans to the Space Age has not received any scholarly attention.

3. For an excellent review of the social and political conditions in South Africa during this period, see Saul Dubow, *Apartheid, 1948–1994* (Oxford: Oxford University Press, 2014).

4. David Harland, *NASA's Moon Program* (New York: Springer, 2010), 174.

5. Tsiao, *"Read You Loud and Clear!,"* 208.

6. Eisenberg, "Reexamining the Global Cold War," 87–88.

7. Charles C. Diggs Jr. papers, Box 194, Folder 1, Moorland-Spingarn Research Center, Howard University, Washington, DC.

8. Diggs's biographical information is based on US House of Representatives, "Diggs, Charles Coles, Jr.," *History, Art & Archives,* http://history.house.gov/People/Detail?id=12254; and Carolyn Dubose, *The Untold Story of Charles Diggs: The Public Figure, The Private Man* (Arlington, VA: Barton Publishing House 1988).

9. Alvin B. Tillery Jr., "Foreign Policy Activism and Power in the House of Representatives: Black Members of Congress and South Africa, 1968–1986," *Studies in American Political Development* 20 (Spring 2006): 89.

10. Robert Gillette, "NASA: Caught between Congress and Apartheid," *Science* New Series 175, no. 4028 (March 24, 1972): 1341.

11. *U.S. Business Involvement in Southern Africa: Hearings, Ninety-Second Congress, First Session, Part 3* (Washington, DC: Government Printing Office, 1973), 198.

12. Ibid., 199.

13. Ibid., 186.

14. Ibid., 184.

15. Gillette, "NASA."

16. Colin G. Garvie, "Hartebeesthoek—50th Anniversary," http://ichthyscybernetics.blogspot.com /2011/03/hartebeesthoek-50th-anniversary.html.

17. Goler T. Butcher, Robert Bates, and Patrick Swygert, "Congress and American Relations with South Africa," *Issue: A Journal of Opinion* 3, no. 4 (Winter 1973): 60.

18. Ibid., 64.

19. Eisenberg, "Reexamining the Global Cold War," 114.

20. Robert Gillette, "South Africa: NASA Inches Out of a Segregated Tracking Station," *Science* New Series 181, no. 4097 (July 27, 1973): 331.

21. Amendment offered by Mr. Rangel, *Congressional Record,* 92nd Cong., 2nd sess., vol. 118, pt. 11 (1972): 13643–13644.

22. Constance Holden, "NASA: Sacking of Top Black Woman Stirs Concern for Equal Employment," *Science* New Series 182, no. 4114 (November 23, 1973): 806.

23. Brenda Gayle Plummer, *In Search of Power: African Americans in the Era of Decolonization, 1956–1974* (Cambridge: Cambridge University Press, 2012), 314.

24. Roger D. Launius, "A Western Mormon in Washington, D.C.: James C. Fletcher, NASA, and the Final Frontier," *Pacific Historical Review* 64, no. 2 (May 1995): 217–241.

25. Holden, "NASA," 807.

26. Jennifer E. Manuse, "The Strategic Evolution of Systems: Principles and Framework with Applications to Space Communication Networks" (PhD diss., MIT, 2009), 35.

27. Cable re NASA Tracking Station, Public Library of US Diplomacy, July 9, 1973, http://www.wikileaks.org/plusd/cables/1973STATE133741_b.html.

28. Michael L. Krenn, *The Impact of Race on U.S. Foreign Policy: A Reader* (New York: Garland, 1999), 180.

PART IV

BROADER CONTEXT

"A Competence Which Should Be Used"

NASA, Social Movements, and Social Problems in the 1970s

CYRUS C. M. MODY

The Apollo 11 Moon landing marked a rather obvious turning point in the American space program. The United States had come from behind to beat the Soviet Union in a declared race to send a piloted craft to the Moon. Now the National Aeronautics and Space Administration (NASA) faced a difficult question: what next? The 1970s were a rudderless period in which NASA struggled to decide among several options for its future. Not entirely coincidentally, the civil rights movement faced a similar set of questions at roughly the same time. In the early 1960s, the civil rights movement won historic but narrowly focused victories in pushing the American legal and political system to more faithfully reflect the tenets of liberal democracy. With those victories won, the movement's social justice objectives took on greater prominence—sometimes in alignment with (and sometimes in competition with) the goals of the era's more nascent social movements: the environmental, antiwar, gay rights, antipsychiatry, and Native American and Latinx civil rights movements, as well as second-wave feminism and the youth counterculture. As Jacquelyn Dowd Hall argues in the essay that gives this volume its title, this "movement of movements" is one of the great achievements of the civil rights movement, though often obscured by undue focus on the movement's "classical phase."[1]

Thus, both the civil rights movement and NASA entered the 1970s with remarkable achievements *in their pasts* and a muddied focus for the future. This chapter argues that this circumstance created the possibility for NASA's *mission* to take on board some of the social justice aims of the social movements of the 1970s. A few concrete steps were indeed taken in that direction. Yet those steps never gained meaningful buy-in from NASA's leadership and were therefore quickly discarded amid the cascading public crises of the 1970s and the neoliberal ascendance post-1980.

Existential Success

One way to think about both NASA and the civil rights movement in the 1970s is that they faced a dynamic of "existential success." NASA's existence pre-1969 was not predicated *entirely* on beating the Soviets to the Moon, but that was the predominant rationale for its rapid growth in funding and personnel in the mid-1960s. Succeeding in that nearly existential mission had two consequences. First, NASA needed a new mission to justify its continued existence. As one assistant administrator argued in 1973, "lacking a 'driver' of the Apollo type, we've got to arrive at an agency rationale."[2] And second, actors outside the space program (other federal agencies, companies, universities, state and local governments) could see that NASA personnel were competent enough to succeed against remarkable odds, *and* that they were not yet attached to a new objective. Those actors therefore attempted to put their own aims on the agendas of the people responsible for beating the Soviets to the Moon.

Similarly, the civil rights movement was not *only* seeking equality before the law pre-1965, but that was the focus of much of its energy and the benchmark by which many defined its success. Passage of the Civil Rights and Voting Rights Acts (and other reforms) seemed to substantially meet that benchmark. De facto, rather than de jure, inequalities in education, housing, medicine, employment, exposure to pollution, and so forth had always been central to the movement's grassroots support, and post-1965 those issues—alongside the injustices of war and colonialism—took on ever greater prominence in civil rights leaders' rhetoric. At the same time, stakeholders in other social movements could see that the practices, rhetoric, and personnel associated with the civil rights movement were effective, and therefore should be co-opted. In the 1970s, social movements that drew on the civil rights template experienced the rapid gains in visibility and efficacy that characterized civil rights a few years earlier.

The consequences of existential success for the civil rights movement are better examined by other, more capable scholars. My point in drawing the parallel between NASA and the civil rights movement is simply to emphasize that NASA's existential success played out in an environment in which there were many social movements wielding the tools of the civil rights movement, and that some people (NASA personnel, figures in the Nixon administration, academic researchers, and so forth) believed those movements could supply the space agency with problems, resources, and legitimacy in return for NASA's supplying those movements with "Space Age" expertise relevant to their aims. Of course, social movements were not the only actors seeking an exchange with NASA; corporations, the military, and nonprofit organizations all looked to the space program's success. Nor did Apollo 11 occasion existential success solely for NASA; aerospace companies faced a similar dynamic and sometimes tried to navigate it in partnership with the space agency.

The exchanges precipitated by existential success took two main forms: an outflow of NASA and aerospace industry personnel to other sectors, and the importation of new R&D projects into NASA and aerospace firms via partnerships with other organizations. This chapter examines the latter, but the outflow of personnel is worth a brief mention and call for further study. Most visible in that flow were former astronauts. The astronaut corps was cut by more than half between 1967 and 1976, mirroring job losses across NASA and the aerospace industry leading to perhaps as high as 20 percent unemployment in the early 1970s.[3] Many ex-astronauts—again, mirroring ex-NASA employees more generally—moved into jobs for which the relevance of their previous experience was somewhat opaque: politician, diplomat, museum director, airline CEO, gubernatorial adviser, vice president of a football team, mining executive, and so forth. More numerous, though, were former NASA and aerospace industry engineers and managers who ended up at companies and especially at government agencies that were trying to apply Cold War R&D, and especially systems engineering, to civilian social problems. The National Science Foundation (NSF), for instance, started a program called Research Applied to National Needs (RANN) in 1969 that by the mid-1970s was absorbing about 10 percent of the NSF's budget to fund research into pollution, energy, disability technologies, mass transit, even garbage collection. The RANN unit was headed by several emigres from NASA.[4] There were similar people sprinkled throughout the upper reaches of the science and engineering community—a NASA diaspora worthy of further research.

From the Moon to the Earth

It was not inevitable that NASA's existential success would lead it to seek partnerships in nonspace domains. It could have been that NASA's new mission simply extended its old one. Many in the agency hoped that Apollo 11 would quickly lead to the "von Braun paradigm" of ongoing Moon missions followed by space stations, lunar colonies, and a mission to Mars.[5] Instead, the Apollo program ended with Apollo 17, and the human spaceflight program shifted—very slowly—to the shuttle. Something similar happened to the planetary probe program. As Peter Westwick has shown, the Jet Propulsion Laboratory (JPL) was allowed to finish off the planetary missions that had entered the pipeline before the 1970s, but the lab faced enormous pressure to find alternative sources of funding once that pipeline dried up.[6]

The political will for NASA to continue on its expansionist trajectory evaporated in the early 1970s. One reason, documented by John Logsdon, was Richard Nixon's instrumental view of NASA. Nixon saw the space program purely in terms of what it could deliver domestically and diplomatically. In terms of diplomacy, Apollo 11 sufficed to humiliate the Soviets, while Apollo-Soyuz (and related minor projects) served as tokens in the ballet of détente. Domestically, Nixon wanted NASA to refocus on terrestrial economic, social, and technological problems, and therefore seriously entertained repurposing it as the "Aeronautics, Space, and Applied Technology Administration" with "government-wide responsibility for the application of technology to national needs."[7] NASA was only one of several organizations that Nixon used in this way. For instance, as the environment became a bigger issue domestically and internationally, Nixon tasked the North Atlantic Treaty Organization (NATO)—NATO!—with creating a Committee on the Challenges of Modern Society to tackle environmental problems.[8]

The American public largely shared Nixon's attitude toward science and engineering. In the late 1960s, campus protestors, members of Congress, practicing scientists and engineers, and ordinary citizens increasingly backed a reorientation of the US R&D system away from national security projects and toward a focus on solving the problems of civil society.[9] Federal funding shifted away from national security R&D and toward agencies with a civilian, applied, interdisciplinary orientation such as the NSF RANN program and the new Environmental Protection Agency and Energy Research and Development Administration. There was a surprising amount of consensus among scientists and engineers as to the social problems they believed politicians wanted them to address. Here, for instance, is a list compiled in 1971 by Rudi Kompfner, a Stanford electrical engineer involved in national discussions on science policy: "Decay of the big cities; Mass transportation; Over-Population; Pollution of the environment; Availability of health care; Natural resources and energy planning; Un- and under-employment; Integration of races, etc."[10] Compare that with another list from 1971 compiled by John Baldeschwieler, a Caltech chemist and Nixon's deputy science adviser: "Urban decay— crime; Drug abuse; Environmental degradation [sic]; Health care; Unemployment; Energy; Education in deprived areas; Productivity and foreign trade; Disease care and prevention; Transportation; Natural disasters."[11] It would hardly be an overinterpretation to see both lists as informed by their authors' (and contemporary politicians') understandings of the civil rights movement and American race relations.

As Roger Launius has shown, NASA was dangerously exposed to shifting public opinion once the race to the Moon was won: from 1970 to 1975 a majority of those polled favored cutting funding for space exploration; at least 40 percent of those polled voiced that opinion through the end of the 1970s.[12] Hence, projects that took NASA away from space and toward the social problems listed above proliferated in the 1970s. Most of these projects were short-lived; thus, while space historians are generally aware of them, they have not received much attention apart from Neil Maher's *Apollo in the Age of Aquarius*.[13] This chapter adds to Maher's analysis by explaining NASA's shift in terms of existential success and by drawing out the commonalities between NASA's situation and the American science and engineering community more generally. In the early 1970s, much of the American public came to believe that the nation's science and engineering community had achieved its existential aim of besting the Soviets on the literal and figurative Cold War battlefield. Now the public and its representatives made it clear that all of the nation's scientists and engineers could and should refocus on solving problems such as pollution and poverty. As America's most visibly capable science and engineering organization, NASA was under particular pressure to respond to that call.

Of course, in some sense NASA's mission had always encompassed contributing to solving terrestrial social problems. The agency's organic charter contained a vague reference to applications of "aeronautical and space science and technology . . . within and outside the atmosphere."[14] In 1962, NASA headquarters interpreted that to justify opening a Technology Utilization Office, which published annual reports enumerating dozens of supposed "spin-offs" of space programs. Yet, as Bruce Seely argues, NASA seems to have measured "utilization" by how widely its documentation circulated rather than by whether its discoveries ended up in civilian technologies.[15]

By the early 1970s, though, headquarters-sponsored "Applications Teams" were "diversif[ying] into other, nonmedical, public-sector areas" such as "air and water pollution, fire safety, housing and urban development, transportation, law enforcement, criminalistics, the postal services, and mine safety."[16] There then seems to have been a fairly rapid transition in the early 1970s from headquarters pushing the field sites to create Application Teams to "the Field Centers themselves submit[ting] proposals based on their own perceptions of the relevance of their in-house skills and capabilities to public-sector problems."[17] At the same time, external actors became more interested in working with NASA on terrestrial applications. For instance, industrial requests to NASA's Technology Utilization program went from ten to twenty per year in the late 1960s to seventy to ninety in the early 1970s.[18] Similarly, reports from this era offered long lists of other government agencies with whom the Technology Utilization program had collaborated: "Department of Health, Education, and Welfare; the Department of Housing and Urban Development; the Bureau of Mines; the Department of Transportation; the Environmental Protection Agency; the Law Enforcement Assistance Association; the National Bureau of Standards; the Veterans Administration and others."[19] This convergence of interests—and eventually action—between NASA and a range of external actors exemplifies the dynamic of existential success.

Finding Partners

An example of how Application Teams enacted that convergence of interests can be seen in a 1974–1975 Johnson Space Center (JSC) test of whether the lunar rover could be adapted for use by people who were quadriplegic. The aim of this and many similar efforts was to resolve a market failure:

> It is unlikely that specialty equipment manufacturers can afford to develop a high-quality, well-engineered product such as the proposed Lunar Rover controller for automobiles. This is an opportunity for NASA, as a public agency responsible for the technology, and the VA [Veterans Administration], as a public agency responsible for the care and service of a large segment of the physically impaired population, to sponsor the development of such a system.[20]

As the quote implies, NASA partnered with the Veterans Administration, as well as with the Department of Transportation and a number of Texas state agencies and non-governmental organizations (NGOs) that worked with people with disabilities. These kinds of partnerships between organizations that represented possible users and those with expertise relevant to terrestrial applications of space technology became quite common in the early 1970s. The focus on disability technologies was also quite typical. Projects to aid people who were blind, who were hard of hearing, or who were paralyzed were very common responses to calls for civilianization of the R&D system in the years around 1970.[21]

Some of these projects aimed to alleviate market failure by ushering a prototype to the point where a private entity could profitably manufacture and sell a commercial version. This mode of working was aided by a new patent policy instituted in 1972 that allowed NASA to grant exclusive licenses on its patents nine months after filing rather than two years after issuance.[22] As in the lunar rover example, these projects often involved quite complex collaborations between NASA, other federal agencies, state governments, academic researchers, and NGOs—with the commercial entity only brought in at the end.

A particularly colorful example of this market-failure approach was a program in the mid-1970s to adapt astronaut food for delivery by mail, volunteer, or social worker to elderly people who were shut in.[23] The program was a result of a request made to NASA by the Texas Governor's

Committee on Aging. Participants included the JSC, the Texas Research Institute of Mental Sciences, United Action for the Elderly (an affiliate of Meals on Wheels), and researchers from the University of Texas (both the School of Public Affairs in Austin and the University of Texas Medical Branch in Galveston). The cost was around a quarter million dollars, about half paid by NASA; the Texas Department of Public Welfare and the Ford Foundation also contributed money. Some of this funding went to an aerospace contractor, Martin Marietta, to develop the "meal system for the elderly" and run field tests.

Programs such as Meal System for the Elderly illustrate how issues and groups that were spotlighted by the civil rights movement were cautiously taken up by NASA. None of NASA's publications about the meal system mention social justice or civil rights. Yet almost half the participants in field trials were beneficiaries of the USDA Food Stamp program; two-thirds were women; 44 percent were African American; and the project report identified another 29 percent as "Mexican-American."[24] Project reports suggested that the same meal system could benefit the non-elderly poor, people with disabilities, and prisoners. All of these were groups whose interests were promoted by the civil rights movement and the new social movements of the late 1960s and 1970s.

Unsurprisingly, though, NASA viewed the concerns of these movements through a paternalist—if not outright patronizing—prism. For instance, NASA suggested the system could be useful to "some urban elderly [who] are within walking distance of stores, but fear to venture out of their homes."[25] A few sentences later in the same report the authors observe that

> some also criticize the food stamp program, because it does not guarantee that recipients will buy food items that contribute to a balanced diet. Such a criticism might be quelled if a balanced, nutritious meal such as that developed by NASA were available for use by food stamp participants.[26]

Nor should it be surprising that commercial manufacturers were less than attentive to the needs of prisoners, the poor, people of color, and/or people with disabilities. In the late 1970s, products spun-off from Meal System for the Elderly were marketed by Oregon Freeze Dry Foods as "Space Age Food, developed for Adults of Retirement Age."[27] Yet as figures 23 and 24

Above: Figure 23. Fieldworker demonstrating the Meal System for the Elderly in the home of a trial participant. From Technology Utilization Program, *Meal System for the Elderly,* JSC-11191 (Houston: Johnson Space Center, 1976), 9.

Left: Figure 24. Publicity shot used in advertising for the Easy Meal line of products sold by Oregon Freeze Dry Foods. The company collaborated on the Meal System for the Elderly, as noted in its ads. The company's participation in the project was also noted by NASA, for instance, in the publication this shot is taken from: Technology Utilization Division, *Spinoff 1978: An Annual Report* (Washington, DC: NASA, 1978), 91.

indicate, Easy Meal's imagined elderly customer was considerably whiter and more middle class than the field testers of the Meal System for the Elderly.

It is relatively easy to understand the relevance of NASA's food technology to terrestrial applications. Indeed, many of NASA's forays into public sector problems in the early 1970s involved straightforward translations of aerospace technology. For instance, NASA partnered with the Federal Aviation Administration to develop better air traffic control technologies and with the Energy Research and Development Administration to create new designs for wind turbines and fuel-efficient airplane engines.[28] In many cases, though, NASA's relevance to solving social problems was indirect and managerial rather than technological. Many people inside and outside the agency believed that what NASA could offer to the solving of social problems was expertise in systems engineering and "space age management."[29] Yet this meant that NASA's engineers were often left to draw somewhat far-fetched analogies between aerospace technology and the social problems they were tasked with solving.

For instance, NASA staff with expertise in biomedicine and communications were deeply involved in a number of efforts to improve the delivery of health care to poor rural communities, especially those with a large proportion of ethnic minorities. These projects were justified on the basis of an explicit analogy between the remoteness and low population density of outer space and, for example, Native American reservations. There is already a cottage industry of histories of one of these projects, STARPAHC, or Space Technology Applied to Remote Papago Advanced Health Care, situated on the lands of the Tohono O'odham Nation in southern Arizona.[30] These studies situate STARPAHC in the history of telemedicine, but not in the context of contemporary calls for the civilianization of NASA and of American R&D more generally. Indeed, one unnoticed clue that STARPAHC should be seen in that light is that it was not unique—NASA participated in several similar efforts in the 1970s in rural or remote areas of New Mexico, Nevada, Alaska, and the Permian Basin region of West Texas.[31] Again, these were all areas whose interests were being vocally represented in the 1970s by the American Indian Movement (AIM), the Chicano movement, and other social justice movements that took inspiration from the civil rights movement.

Leveraging the Urban Crisis

Both Meal System for the Elderly and STARPAHC (and other telemedicine projects) were targeted primarily at rural areas. Yet urban problems were of at least equal concern within NASA for most of the 1970s. In fact, many of the publications issued by NASA's public sector projects pictured urban and rural areas as similarly amenable to interventions founded on expertise in aerospace technology. Sometimes this supposed similarity led to arguments that solutions to rural problems would also find application in cities and vice versa (as with Meal System for the Elderly). As Sam Pool (one of the leaders of STARPAHC) and colleagues put it, "consider groups who live in remote areas or, for that matter, in the inner cities who should have access to local health services and to the health care establishment."[32] At other times, the concentration of poverty in rural and urban areas was taken to mean city and countryside would compete for scarce resources. As the report on a remote health care project in New Mexico put it, "More is heard of ghetto problems, but rural health care has suffered particularly because most young physicians . . . and their wives prefer to settle in urban areas. . . . Even ghetto practice is often near splendid medical centers."[33]

The sentiment is grating, but the author was correct that public discourse concerning urban problems was mounting rapidly in this era. Note the first items on Kompfner's and Baldeschwieler's lists of social problems mentioned above; these are very much in line with figure 25's Google Ngram for mentions of "urban crisis/urban problems" versus those for "rural crisis/rural problems." Of course, to return to the theme of this volume, there should be no automatic link between "urban problems" and civil rights. Indeed, NASA publications generally avoided mentioning race in discussing urban (or rural) issues. And yet the racial subtext of talk about an "urban crisis" was never far away. For instance, in a 1970 pamphlet issued by the Aerospace Industries Association (AIA) titled *Aerospace Technology: Creating Social Progress*, "urban affairs" were the first site of "social progress" presented—and also the only site (of seven) in which the accompanying illustration depicted people of color.[34]

Nixon's America, though, saw considerable disagreement as to *how* urban problems were connected to race and to the civil rights and other contemporary social movements. From the point of view of many in the New Left, the civil rights movement, and the latter's splinter groups such as the Black Panthers, the urban crisis was caused by institutions—including the

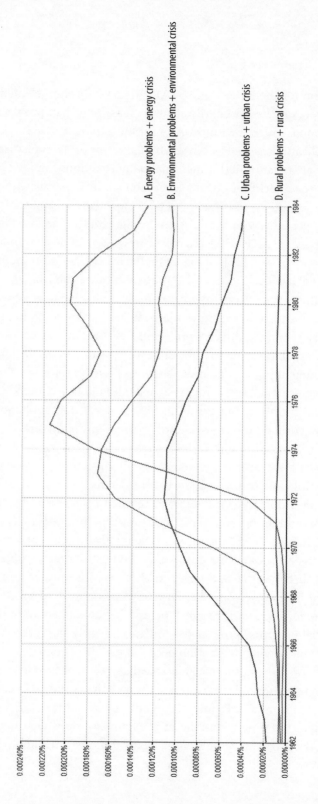

Figure 25. Google Ngram showing the prevalence of English-language mentions of problem/crisis in four domains: rural, urban, environment, and energy. Smoothing of one, and Ngram compiled on January 23, 2019. Note the successive peaks: urban (~1972), environment (~1973), energy (~1975 and again ~1980). Note also the low level of interest in rural problems relative to the other domains.

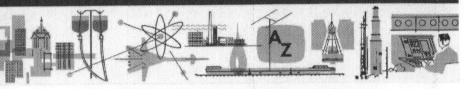

Figure 26. Cover and page 2 of *Aerospace Technology: Creating Social Progress* (Washington, DC: Aerospace Industries Association, undated but probably 1970). Multiple versions of this pamphlet exist with the same publication information; this one was located in the Johnson Space Center History Collection, University of Houston Clear Lake Archives and Special Collections, Center Series, Poindexter Reference Files, Section 89, Box 6, Folder "Space Benefits / Technology Utilization–General Information." Image courtesy of Aerospace Industries Association.

institutions of American science and engineering such as NASA—that had a vested interest in bringing about, or at least doing nothing to reverse, the second-class citizenship of most city dwellers, especially those of color.[35] On the other hand, many Nixon voters no doubt believed the urban crisis stemmed from what they saw as growing disregard for authority among African Americans, young people, and the antiwar Left.[36]

This difference in perspective led in turn to differing opinions as to *what* urban problems scientists and engineers should try to solve. From the point of view of NASA, though, differing explanations for the urban crisis simply meant an even wider array of research topics by which the agency could make itself relevant to the pressing social issues of the day. Thus, on the one hand, NASA (particularly the JPL) and aerospace firms were deeply implicated in Nixon's "law and order" rhetoric aimed at harvesting the votes of white Democrats who disliked both the civil rights and antiwar movements. Notably, the second illustration in the AIA's 1970 pamphlet showed two police officers surrounded by banks of computers under the caption "Information Systems." The JPL had projects to help local law enforcement agencies communicate better with each other (using NASA's communications networking expertise) and to develop fingerprint recognition software for the Federal Bureau of Investigation (FBI) (using image processing software originally developed for the planetary probe program).[37] In 1974, JPL researchers even invented an Automated Drug Identification device to detect "drugs in blood or urine samples taken from suspected narcotics users" at the request of the Los Angeles Police Department (LAPD).[38] The short-lived NASA/New York City Applications Project came up with a similar "portable device for detecting extraordinarily small amounts of heroin in a urine sample."[39]

On the other hand, the Nixon administration also used NASA to position itself relative to the "urban crisis" in a way that would appeal to the Nelson Rockefeller and George Romney wing of the Republican Party. The JPL, for instance, ran a multiyear Four Cities Program in which the city managers of four municipalities in California (Anaheim, Pasadena, San Jose, and Fresno) were each teamed with a JPL engineer and an engineer from an aerospace firm (TRW, Lockheed, Aerojet, and Northrup) to see how aerospace and systems engineering might be applied to their city's issues.[40] The AIA similarly bragged of its members' efforts in "school community relations . . . job training . . . city programming and budgeting . . . [and] justice, welfare studies."[41] As Jennifer Light has documented,

NASA participated in numerous—though generally rather small—urban studies projects in the late 1960s and early 1970s.[42]

At the JSC, urban applications actually made their way onto the organization chart in the form of an Urban Systems Projects Office (USPO). The USPO was founded in 1972 with a staff of thirty-five, including both civil servants and contract personnel from Lockheed and Boeing.[43] The unit was NASA's stake in a project headed by the Department of Housing and Urban Development (HUD) that also included the Environmental Protection Agency and Oak Ridge National Laboratory (part of the Atomic Energy Commission). Notably, the NASA-HUD collaboration seems to have come about in part because Harold Finger, the assistant secretary for research and technology, had been poached from NASA by HUD secretary Romney. The impetus for the USPO's main project, the Modular Integrated Utilities System (MIUS), however, was the Nixon administration's high-profile New Technology Opportunities (NTO) program. The aim of the program was to make work for the thousands of aerospace engineers laid off due to the contraction in both NASA's and the military's procurement budgets by establishing partnerships among aerospace firms, national security agencies, and civilian agencies to make them more efficient and technology oriented.

The arc of the MIUS project stands in for the arc of most public sector programs at NASA in the 1970s. As a 1981 report on MIUS put it, "After about 6 months of intensive design work, the USPO team made its first presentation to HUD, and was shocked to find out that HUD did not intend to use advanced technology but, instead, planned to build MIUS systems using only existing equipment. HUD set a new rule for the project—only bondable catalog available hardware could be used in the MIUS development."[44] That is, HUD was asking a group of aerospace engineers who knew a lot about designing one-of-a-kind space vehicles to instead design terrestrial housing using other people's mass-produced equipment. Apparently, some wanted the JSC to get out of the project at that point. The 1981 study quotes the project manager saying he decided to keep going because MIUS was an exciting "challenge," but I think equally important is that there wasn't an alternative—in 1972 there wasn't a space vehicle for these engineers to work on.

The end result was, essentially, an advanced cogeneration plant for a small planned community. This was not an ideal outcome for either HUD or NASA. For JSC engineers, the scheme was too *low-tech*, and they continued adding in higher-tech elements, designed-from-scratch technologies to supplement the plans they delivered to HUD. But for HUD, even the most basic design was too *high-tech*; the agency would never be able to get developers to include it without a government subsidy, which HUD was never in a position to supply. For JSC director Chris Kraft, the program had been useful in the early phase when it had powerful sponsors in the Nixon administration and not much else was going on. But by 1974 those sponsors, William Magruder and Edward David, were out, and the shuttle program was beginning to heat up.

Thus, Kraft abruptly canceled future plans and brought the project to a close in 1976. As USPO deputy manager Harold Benson reported, Kraft reminded project personnel that

> they were working for the Manned Spacecraft Center and that what they were working on was not in any way related to manned space. Dr. Kraft went on to say that the JSC has the best engineers in the world, and the world is full of problems; if he would allow his engineers to go to work on all the world's problems, they would fragment the agency and the NASA's mission.[45]

By that logic, though, the USPO should never have been formed. Yet it was formed and received resources and personnel so long as urban problems were a major focus of the Nixon administration and so long as the shuttle was only in the planning stages at the JSC. In 1970–1971, as Apollo was winding down, much of the American public and government believed that "NASA has a competence which should be used to complement HUD's activities in developing new housing."[46] It was only later that a project aimed at improving multifamily housing no longer seemed desirable within America's *space* agency.

Conclusion

As the USPO case shows, though, NASA's interest in urban problems— and, more generally, public sector problems of the type encompassed by Rudi Kompfner's and John Baldeschwieler's lists above—attenuated over the course of the 1970s and virtually disappeared in the 1980s. The reasons are manifold. Certainly, the return to active missions and growing budgets for both human and robotic spaceflight were part of it. The neoliberal turn, first under Jimmy Carter and even more under Ronald

Reagan, also contributed; increasingly, NASA was tasked with assisting the privatization of space, rather than with ameliorating the public sector.[47] But another reason, surely, is that the civil rights movement, the environmental movement, the New Left, and other early 1970s social movements lost traction in American political discourse and were unable to prevent that discourse from refocusing on other issues. In particular, the "energy crisis"—which preceded but was brought into sharp relief by the 1973 Organization of Arab Petroleum Exporting Countries (OAPEC) embargo—shifted a great deal of attention away from urban and environmental problems and the challenges of people of color, people of limited means, and people with disabilities (see figure 27).

At the JPL, especially, the energy crisis quickly supplanted other public sector problems and then facilitated the disappearance of public sector topics entirely. In 1975, the JPL concluded a memorandum of understanding with the Energy Research and Development Administration (and later with its successor, the Department of Energy), leading to a steady flow of contracts to evaluate the fuel efficiencies of various engines, to develop different systems for generating electricity from solar energy, and so forth.[48] By 1978, almost 20 percent of JPL personnel and about 15 percent of the lab's funding were dedicated to "Energy and Technology Applications."[49] A few of these projects had some connection to the JPL's earlier interest in urban problems: research in personal rapid transit systems or into the aerodynamics of subway trains moving through tunnels, for instance.[50] But most were quite divorced from the research agenda promoted by the civil rights and accompanying social movements. A few of the JPL's energy projects even aimed to assist populations associated with antipathy to the civil rights movement. For instance, the JPL sponsored a multidisciplinary study of coal mining that combined social scientific research on mining communities with technological explorations of how to mine coal more efficiently.[51] Similarly, the JPL's most resource-intensive energy project was probably a group of solar collectors—constructed near its Barstow, California, Deep Space communications installation—that tested designs for solar power installations for small farming communities.[52]

Nevertheless, the Reagan administration slashed budgets for solar R&D across all agencies, including NASA, and energy work quickly disappeared from the JPL. Overall, NASA's interest in the public sector waned quickly in the late 1970s and almost vanished in the 1980s. So what should we make of this brief period when a sizable number of NASA scientists

Figure 27. Artist's conception of a solar power installation "supplying electricity to a rural community." From *Mirror on the Sun*, JPL400-10 (Pasadena, CA: Jet Propulsion Laboratory, 1980).

and engineers threw themselves into work that complemented—but also at times directly opposed—the agenda of the civil rights and accompanying social movements?

My hope is that this chapter makes two contributions to space history. First, it is important to acknowledge the paths not taken in NASA's history. In the 1970s, there was the very real, if not wholly likely, possibility that energy or "civil systems" R&D would come to occupy some large proportion of the agency's work. Second, even if space will nevertheless always be the main focus of space history, there are aspects of NASA's explicitly space-oriented activities in the 1970s that are hard to understand if you don't include this penumbra of activities in the wider public sector. For instance, the Landsat remote sensing program is once more becoming a topic of interest among space historians—in part as a marker of NASA's long-term interest in environmental topics and in part as token of NASA's increasing orientation to the market.[53] However, if one looks at how Landsat and NASA's other remote sensing technologies were used in the 1970s, public health—including urban public health—took up a surprising amount of the research portfolio.[54] Projects that aimed to aid Native Americans and Alaska Natives, too, were used to advertise Landsat's benefits.[55] Indeed, in 1972 leadership of the Landsat program was assigned to the JSC; that seems an oddity if the JSC's main mission is taken to be "human spaceflight," but is understandable at a time when the JSC's mission *might* have broadened to human social problems more generally.[56]

A slightly different way of putting my point is this: the other chapters in this volume are largely concerned with issues relating to personnel, policy, and community relations. One might call these topics the "who" and "where" of NASA's relationship to the civil rights movement. This chapter, by contrast, is mostly concerned with the "what"—that is, with the ways that NASA's mission and technical content did or did not internalize the civil rights movement's vision for a more responsive and just America. As Blount and Molina show in this volume, the "what" of NASA was indeed an important focus of critique by leading figures in the civil rights movement. As the other chapters in this volume argue, pressure from the civil rights and accompanying social movements led to partial reforms to NASA's personnel and community relations policies. At the beginning of the 1970s, similar pressure facilitated similar partial changes in NASA's mission—only to see those changes disappear by the end of the decade. Either NASA's mission has been less amenable to enduring reform than its personnel policies or perhaps changes in personnel policies have simply papered over a more intractable organizational conservatism.

Notes

1. Jo Freeman and Victoria Johnson, eds., *Social Movements since the Sixties* (Oxford: Rowman & Littlefield, 1999), especially Douglas McAdam, "The Decline of the Civil Rights Movement," 325–348; Jacquelyn Dowd Hall, "The Long Civil Rights Movement and the Political Uses of the Past," *Journal of American History* 91, no. 4 (2005): 1233–1263.

2. John Donnelly memo to James Fletcher quoted in Matthew H. Hersch, *Inventing the American Astronaut* (New York: Palgrave Macmillan, 2012), 138.

3. Hersch, *Inventing the American Astronaut*, 104; Matthew Wisnioski, *Engineers for Change: Competing Visions for Technology in 1960s America* (Cambridge, MA: MIT Press, 2012), 36.

4. Richard J. Green and Wil Lepkowski, "A Forgotten Model for Purposeful Science," *Issues in Science and Technology* 22, no. 2 (2006): 69–73.

5. Michael J. Neufeld, "The 'Von Braun Paradigm' and NASA's Long-Term Planning for Human Spaceflight," in *NASA's First 50 Years: Historical Perspectives*, ed. Steven J. Dick (Washington, DC: NASA, 2009), 325–347.

6. Peter J. Westwick, *Into the Black: JPL and the American Space Program, 1976–2004* (New Haven, CT: Yale University Press, 2007).

7. John Logsdon, *After Apollo: Richard Nixon and the American Space Program* (New York: Palgrave MacMillan, 2015), 182–183.

8. Jacob Darwin Hamblin, *Arming Mother Nature: The Birth of Catastrophic Environmentalism* (Oxford: Oxford University Press, 2013), 192.

9. Eric J. Vettel, *Biotech: The Countercultural Origins of an Industry* (Philadelphia: University of Pennsylvania Press, 2006); Stuart W. Leslie, *The Cold War and American Science: The Military-Industrial-Academic Complex at MIT and Stanford* (New York: Columbia University Press, 1993); Kelly Moore, *Disrupting Science: Social Movements, American Scientists, and the Politics of the Military, 1945–1975* (Princeton, NJ: Princeton University Press, 2008).

10. Unpublished thought piece "Generalities about a national science policy," Rudi Kompfner, undated but probably 1971, Box 1, Folder 1, collection SC 194 ACCN 85–074, Rudolf Kompfner papers, Special Collections, Stanford University.

11. Memo, John Baldeschwieler to Hu Heffner, July 9, 1971, Box 114, Folder "Domestic Council Study Group on NTO," Edward David Papers, Nixon Presidential Library.

12. Roger D. Launius, "Public Opinion Polls and Perceptions of U.S. Human Spaceflight," *Space Policy* 19, no. 3 (2003): 163–175.

13. Neil M. Maher, *Apollo in the Age of Aquarius* (Cambridge, MA: Harvard University Press, 2017), 43–60.

14. National Aeronautics and Space Act of 1958, Public Law 85-568, 72 Stat., 426–438, July 29, 1958.

15. Bruce Seely, "NASA and Technology Transfer in Historical Perspective," *Comparative Technology Transfer* 6, no. 1 (2008): 1–16.

16. "Technology Applications Progress Report for the Period 1 January 1972–31 December 1972," NASA Technology Utilization Office, December 1972.

17. Ibid.

18. "Technology Utilization Program Report," December 1973, NASA SP-5119.

19. Maher, *Apollo in the Age of Aquarius*, 2.

20. Sam McFarland, "Application of Features of the NASA Lunar Rover to Vehicle Control for Paralyzed Drivers," Final Report for contract NAS 9-14473, August 1975, 67.

21. Cyrus C. M. Mody, "'An Electro-Historical Focus with Real Interdisciplinary Appeal': Interdisciplinarity at Vietnam-Era Stanford," in *Investigating Interdisciplinary Research: Theory and Practice across Disciplines*, ed. Scott Frickel, Barbara Prainsack, and Mathieu Albert (New Brunswick, NJ: Rutgers University Press, 2016), 173–193; Cyrus C. M. Mody, "Santa Barbara Physicists in the Vietnam Era," in *Groovy Science: The Counter-Cultures and Scientific Life, 1955–1975*, ed. David Kaiser and W. Patrick McCray (Chicago: University of Chicago Press, 2016), 70–107.

22. Maher, *Apollo in the Age of Aquarius*, 2.

23. *Meal System for the Elderly: Conventional Food in Novel Form*, Lyndon B. Johnson School of Public Affairs Policy Research Project Report No. 16, 1977.

24. Ibid., 9.

25. Jurgen Schmandt, ed., *Shelf Stable Meals for Public Sector Uses: Report of a Conference on the NASA Meal System* (Austin, TX: Lyndon B. Johnson School of Public Affairs, 1977), 36.

26. Ibid.

27. Marketing brochure, Oregon Freeze Dry Foods, Inc., undated, Box 7, Folder "Meal Systems for the Elderly 1971–1977," Center Series, Section 89, Poindexter Reference Files, Johnson Space Center History Collection, UHCL Archives, Alfred R. Neumann Library, University of Houston Clear Lake.

28. Mark D. Bowles, *The "Apollo" of Aeronautics: NASA's Aircraft Energy Efficiency Program, 1973–1987* (Washington, DC: NASA, 2010).

29. Roger D. Launius, "Managing the Unmanageable: Apollo, Space Age Management, and American Social Problems," *Space Policy* 24 (2008): 158–165.

30. Lisa Cartwright, "Reach Out and Heal Someone: Telemedicine and the Globalization of Health Care," *Health* 4, no. 3 (2000): 347–377; Rashid L. Bashshur and Gary W. Shannon, *History of Telemedicine: Evolution, Context, and Transformation* (New Rochelle, NY: Mary Ann Liebert, 2007), 195–202.

31. National Academy of Engineering Subcommittee on Technology and Systems Transfer, "Study of Some Aspects of Aerospace Technology Utilization," April 1, 1972, Box LS84, Folder XX0D00506, 20; Permian Basin Regional Planning Commission, "Application for Grant for Permian Basin Region Emergency Medical Services," March 31, 1975, Box LS79, Folder XX0D00742, I-1, both in NASA Life Sciences Collection, McGovern Historical Center, Texas Medical Center Library.

32. R. S. Johnston, George G. Armstrong, and Sam L. Pool, "A Concept for an Expanded Health Services System," ca. October 29, 1971, Box LS84, Folder XX0D00548, NASA Life Sciences Collection, McGovern Historical Center, Texas Medical Center Library.

33. Final report, "A Remote Area Health Services Research, Demonstration, and Evaluation Project," vol. 2, "Implementation Plan," contract HSM 110–69–243, State of New Mexico Health and Human Services Department, Box LS84, Folder XX0D00346, NASA Life Sciences Collection, McGovern Historical Center, Texas Medical Center Library.

34. *Aerospace Technology: Creating Social Progress* (Washington, DC: Aerospace Industries Association, 1970).

35. Alondra Nelson, *Body and Soul: The Black Panther Party and the Fight against Medical Discrimination* (Minneapolis: University of Minnesota Press, 2011).

36. Rick Perlstein, *Nixonland: The Rise of a President and the Fracturing of America* (New York: Scribner, 2008).

37. R. L. Sohn et al., "Application of Mobile Digital Communications in Law Enforcement," JPL-SP-43-6-REV-1, 1974; B. D. L. Mulhall, "FBI Fingerprint Identification Automation Study," JPL Publication 80-79, November 15, 1980. There also appears to have been a bullet ballistics identification project in 1974; Technology Utilization Program Report 1974 (Washington, DC: NASA, 1975), 52.

38. James J. Haggerty, "Aid for the Medical Laboratory," in *Spinoff 1986* (Washington, DC: NASA, 1986), 60–61.

39. John Darnton, "NASA Consultant to the City Is Finding Down-to-Earth Uses for Space Technology," *New York Times*, October 13, 1973.

40. H. L. Macomber and James H. Wilson, "California Four Cities Program, 1971–1973," JPL-SP-43-4, May 15, 1974.

41. National Academy of Engineering Subcommittee on Technology and Systems Transfer, "Study of Some Aspects of Aerospace Technology Utilization," 20; Permian Basin Regional Planning Commission, "Application for Grant for Permian Basin Region Emergency Medical Services," I-1.

42. Jennifer S. Light, *From Warfare to Welfare: Defense Intellectuals and Urban Problems in Cold War America* (Baltimore: Johns Hopkins University Press, 2003), 136–160.

43. There are several folders of contemporary documents relating to MIUS at the Johnson Space Center collection at the University of Houston Clear Lake and in the Edward David papers at the Nixon Presidential Library, as well as a good post hoc summation from which I take most of the details in the next paragraph: Lyn Gordon-Winkler, "NASA and MIUS—A Study in Bureaucratic Failure," Government 5534.01 [probably a course number, possibly at the University of Houston], December 11, 1981, Center Series, Section 97, Box 2, Folder 10 "Winkler 1981," Johnson Space Center History Collection, UHCL Archives, Alfred R. Neumann Library, University of Houston Clear Lake.

44. Gordon-Winkler, "NASA and MIUS—A Study in Bureaucratic Failure."

45. Ibid., 14.

46. "Evaluation of the Housing Development Program According to Domestic Council Criteria," undated but must be 1971, Box LS105A, Folder XX0D00266, NASA Life Sciences Collection, McGovern Historical Center, Texas Medical Center Library.

47. Andrew J. Butrica, "The Reagan White House and the Conservative Agenda for Space," *White House Studies* 8, no. 4 (2008): 501–516.

48. For example, R. Rhoads Stephenson, "Should We Have a New Engine? An Automobile Power Systems Evaluation," JPL SP 43-17, August 1975; M. L. Peelgren, "Salton Sea Project, Phase 1," DOE/JPL-1060-44, January 15, 1982.

49. "Supporting Material for Executive Council Retreat, 1979," November 20, 1979, Folder 2, JPL 229, JPL Executive Council Retreat papers, JPL Archives.

50. Charles F. Lockerby, "Obstacle Detectors for Automated Transit Vehicles: A Technoeconomic and Market Analysis," NASA-CR-164661, December 1979; Donald W. Kurtz and Bain Dayman Jr., "Experimental Aerodynamic Characteristics of Vehicles Traveling in Tubes," Technical Memorandum 33-731, July 15, 1975.

51. William B. Mabe, "Economic Baselines for Current Underground Coal Mining Technology," JPL 79-122, December 15, 1979.

52. "Small Community Solar Thermal Power Experiment," JPL 400-82/3, November 1980.

53. Roger D. Launius, "'We Will Learn More about the Earth by Leaving It Than by Remaining on It': NASA and the Forming of an Earth Science Discipline in the 1960s," in *Innovation in Science and Organizational Renewal*, ed. Thomas Heinze and Richard Münch (New York: Palgrave-Macmillan, 2016), 211–242; Brian Jirout, "One Space Age

Development for the World: The American Landsat Civil Remote Sensing Program in Use, 1953–2003" (PhD diss., Georgia Institute of Technology, 2016).

54. For example, M. Rush and S. Vernon, "Remote Sensing for Urban Public Health," *Photogrammetric Engineering and Remote Sensing* 41 (1975): 1148–1155; C. M. Barnes et al., "Applications of Remote Sensing in Public Health," in *International Symposium on Remote Sensing of Environment, October 2–6, 1972* (Ann Arbor: Environmental Research Institute of Michigan, 1973), 677–684.

55. Howard Allaway (NASA News), "Satellite Pictures Help Alaskan Natives Choose Best Land," Release No. 76–63, April 1, 1976; Dick McCormack and Mack Herring (NASA News), "Landsat Satellite to Inventory Navajo Nation's Resources," Release No. 77–223, October 18, 1977.

56. Pamela E. Mack, *Viewing the Earth: The Social Construction of the Landsat Satellite System* (Cambridge, MA: MIT Press, 1990), 97.

The Gates of Opportunity

NASA, Black Activism, and Educational Access

ERIC FENRICH

The Civil Rights Act of 1964 left many Americans asking whether the deliberate efforts to include underrepresented groups through affirmative action measures were necessary since existing civil rights laws already prohibited discrimination. Speaking at Howard University on June 4, 1965, President Lyndon Baines Johnson rationalized the concept by arguing that merely eliminating overt discrimination was insufficient and that African Americans must receive education and jobs that led to greater prospects. "It is not enough just to open the gates of opportunity," he explained, "all our citizens must have the ability to walk through those gates."[1]

LBJ believed one solution to American social problems would come from space. As early as 1957, following the launch of *Sputnik*, Johnson and his advisers equated the conquest of space with the advancement of social justice and social security. Accordingly, Johnson believed that the exploration of space would feed the American economy, even arguing, as president, that the National Aeronautics and Space Administration (NASA) was a part of his Great Society initiatives since it infused poor southern communities with federal investment in high technology.[2] NASA was willing to open the gates, but African Americans could not get through due to racism as well as the lack of educational opportunities.

Education represented a means of participation, first, in the training of engineers and technicians who would contribute to the US space effort and, second, in allowing African Americans to gain the necessary knowledge for full participation in American society. White students had greater access to the education needed to study math and science. Black students, in contrast, had severely limited options with only a handful of black colleges offering classes in the necessary fields. More so, most black students lacked the essential education to qualify for *any* college.

Sputnik and the integration of Central High School in Little Rock, Arkansas, allowed black activists to appropriate the events and language contemporary to the Cold War as an additional means of improving access and creating true equality. These groups would seek out or create programs designed to improve educational access and to assist with job training to produce a pool of qualified students and employees. Getting the gates open represented only part of the solution; civil rights advocates also sought to provide the necessary capabilities needed to walk through them.

The October 4, 1957, launch of the Soviet-built *Sputnik* challenged the United States' claim as the world's industrial and technological leader, one month after the highly charged racial integration of Central High School in Little Rock. In addition to the political and technological race initiated by *Sputnik*, a comparison of the Cold War rivals' educational systems became a focal point, particularly in the fields of mathematical and scientific studies. Politicians clambered to find out how the Soviet Union had surprised—and surpassed—the United States in technological capabilities. These politicians demanded that the United States strengthen its educational system, particularly in mathematics and the sciences, in order to regain technological superiority over the nation's Cold War rival.

Many black activists insisted that the American social system and the lack of quality educational access for minority students stood as the reason for the American failure to launch a satellite before the Soviets. The persistence of racial bigotry signified not only neglect by the federal government for failing to provide opportunity for all its citizens but also represented a national danger as the government willfully chose to ignore the potential of its own people, particularly by adhering to the "old thinking" of racial prejudice.[3] James L. Hicks, editor of the *New York Amsterdam News*, pointed out that "thousands of Negro kids . . . willing to help out our government in this crisis" were not getting the opportunity. He noted that three-fourths of the nation's twenty million African Americans lived in the South: "That's exactly 15,000,000 potential engineers which you can immediately write off because of racial prejudice."[4]

In January 1957, just over eight months prior to *Sputnik*, Julius A. Thomas, a National Urban League (NUL) investigator, proposed a new project designed to overcome the reluctance of African American students to pursue studies in the fields of science and engineering. He intended to enroll at least 1,000 young people into what he termed "Future Scientists of America," a coordinated effort with school principals, counselors, and teachers.[5] Believing that students largely reflected the lack of knowledge and the indifference shared by parents regarding these possibilities, the NUL stipulated that a key facet of this new program would be the "stimulation of parents" and other influential people in the community. Without the faith, guidance, and material help needed from adults, many of these young people would do nothing more than "shop the windows of opportunity" because they did not believe in themselves.[6]

Although the formation of the program began nearly a year before the Soviet satellite orbited Earth, the NUL seized on *Sputnik* as a means of promotion. Citing the recent developments of "national and international significance," the NUL announced in February 1958 its preparation of a nationwide program—now called Tomorrow's Scientists and Technicians (TST)—to encourage young people to choose careers in the sciences in response to the tragic shortage of trained professionals in those "critical fields that are necessary for the national security effort."[7] According to the NUL's data, the proportion of black males in 1950 that worked in professional and related fields was approximately one-fourth the proportion among white men. Fewer than 200 of the 28,000 engineers who graduated in 1957 were people of color. To keep the issue in a Cold War context, the NUL added that this deficiency contributed to the downfall of the American international image: "Here, then, is the waste of human resources that has shamed us in the face of the Soviet educational challenge."[8]

Officially launched in June 1958, the TST program pursued a first-year goal of enrolling 1,000 students in college courses for the purposes of being trained as scientists and engineers, with local branches of the NUL in sixty-three cities organizing the program.[9] During the program's first year, thirty NUL cities reported contact with local schools, the creation of advisory and sponsoring committees, and, in some cases, the establishment of TST Career Clubs. By 1960, total enrollment reached approximately 2,000 young people, with 400 adult advisers.[10] Activities included field trips to laboratories, engineering businesses, and NASA facilities with guest speakers, including scientists, technicians, and engineers from the space agency and other local companies.[11]

While the TST program aimed to provide black youth with role models, the organization's early success was apparently mixed. TST documents show increases in attendance, an increased interest in math and the sciences, and an eagerness for more from its youthful participants. Unfortunately, documents through 1961 also demonstrate a continued reluctance by parents to encourage their children in educational advancement and training because of the ongoing fear that it would be "wasted" because of limited opportunities.[12] Although the TST program attempted to counter the problem, the paucity of relatable role models left parents wary of giving their children false hopes. Past experience suggested that it was futile to pursue advanced education for job opportunities that would not be there or would be so limited in number that it was unrealistic to devote significant effort toward them.[13]

On May 25, 1961, when President John F. Kennedy called upon the "entire nation" to participate in the venture to the Moon, the question was whether "everyone" would be allowed to participate.[14] During NASA's early years, management "walked a tightrope" attempting to balance the nation's desire to compete technologically with the Soviet Union and the changing attitudes toward social equality.[15] In its management manual, the agency sought to establish procedures in accordance with President Dwight D. Eisenhower's Executive Order 10590, which stipulated that the "head of each Federal agency is responsible for effectuating the policy with respect to all personnel matters under his jurisdiction."[16] Accordingly, NASA administrator James Webb "personally" assumed responsibility in April 1961 for the space agency as a whole while informing his program directors and staff officers that they "should take the initiative" in their respective areas of concern.[17] The results of a government-wide survey at the end of the year demonstrated that black employment at NASA equaled 1.7 percent of its total workforce, one of the lowest ratios among government agencies, compared, for example, with 8.3 percent in cabinet agencies and 13.9 percent in independent agencies. In response, Webb requested that the agency's recruiting officers "seek qualified Negroes" to fill open positions and specified that "they must be offered employment when they are equally well-qualified . . . [and] given equal opportunity for training and advancement after they are employed."[18]

NASA faced more of a challenge with the former request—finding qualified black candidates—than with the latter—training and advancement of those already hired. Despite an increased American willingness to hire black professionals, the educational system had yet to provide a significant number of potential employees in the four years since *Sputnik*. James Hicks connected the lack of equal educational access with the failure of the American space program to keep up with its Russian counterpart. While Russian engineers sent a man into Earth-orbit, American racism prevented black engineering students from attending any of the accredited engineering schools south of Washington, DC.[19]

At the end of 1961, Paul Bikle, the director of the Flight Research Center, responded to Webb's request, first, by stating that "the emphasis placed on the few negroes [*sic*] employed in professional type positions is not too realistic" because of the lack of equal educational opportunities they had received. Until that situation could be corrected, which he believed would take several years, "we cannot all of a sudden expect great numbers of Negroes in professional positions when they had not been prepared to assume these responsibilities." Bikle suggested that increased pressure and legislative action on educational institutions to accept and provide adequate curriculum would be necessary to increase the limited supply of black professionals. "We asked industry," he concluded, "to place negroes [*sic*] in professional positions yet the resources are limited. It appears we are endeavoring to stimulate a demand, but until it provides a supply, the demand factor will gain little momentum."[20]

During the early years of the space program, the black press—African American–owned media—made its contribution to providing role models with stories of African Americans working directly and indirectly with the American space program. Along with promoting racial pride, the articles demonstrated that, contrary to popular perceptions, black individuals possessed the capabilities to succeed in the mathematical and technical fields. Shortly after Alan Shepard became the first American in space in May 1961, the *New York Amsterdam News* featured a front-page story detailing the contribution of Katherine G. Johnson to his flight. Her paper, cowritten with Ted Skopinsky, was a "key document" that provided a "definite method" by which an astronaut's and/or a satellite's position could be projected. In addition to Johnson, eight other individuals warranted attention, ranging from truck drivers at the Cape, theoretical mathematicians, and an air technical intelligence officer "whose work is so secret that nobody . . . would talk about what he is doing."[21] In each case, an emphasis was placed on their educational backgrounds and the progressive increases in their respective salaries. Their perseverance emphasized the necessity of obtaining the best possible education to qualify for the vast field of new job opportunities.[22] Even so, by 1964 the recently renamed John F. Kennedy Space Center reported that only three of the six leading Historically Black Colleges and Universities (HBCUs) visited by their personnel indicated an interest in the prescribed curriculum. None had taken the necessary steps to add the coursework.[23]

Ten years after its inception, NASA continued to struggle in its attempts to integrate its workforce, hindered by the decades of educational neglect of American minorities. Throughout the 1960s, NASA's appointment of blacks hovered at approximately 3 percent of its total workforce. By 1970, however, the number of African Americans holding positions as managers and professionals registered at less than 1 percent in each category, with the majority still working in clerical or unskilled labor jobs.[24] The gates of opportunity were slowly opening, but there still appeared to be a significant reluctance at attempting to walk through.

In March 1971, the agency hired a new administrator, James C. Fletcher, with George M. Low as his deputy. The two men sought to improve NASA's poor standing related to diversity. On September 1, 1971, NASA issued a special announcement regarding the establishment of the Equal Employment Opportunity (EEO) Office within the agency. Fletcher announced the hiring of Ruth Bates Harris, a fifty-two-year-old African American woman, as NASA's new director of the EEO Office and director of contract compliance.[25] In her new capacity within NASA, she would oversee the formulation and implementation of policy as well as compliance by the agency's contractors. Less than a week later, however, NASA rescinded the offer and tendered her the deputy director's position, giving the primary position to a white male already working at the agency. Despite being demoted before she had even started, Harris accepted the lower position at NASA headquarters, believing that she could "help the agency be responsive to those critical social problems awaiting solutions back on earth."[26]

Eighteen months later, in April 1973, Harris received a promotion to the position of deputy assistant administrator in the EEO Office. In order to realize a cooperative effort within NASA's workforce, Harris and two other senior aides, Joseph Hogan and Samuel Lynn, submitted a report providing documented proof to Fletcher of how some NASA managers showed only a passing interest in the administrator's guidelines and, while providing all the necessary rhetoric regarding equal opportunity hiring, failed to act upon it. The trio had two purposes in issuing the report: first, to provide a history of past failures within NASA's EEO Program and, second, to offer a blueprint for the "highly successful program which it could and should have." The forty-page report, dated September 20, 1973, referred to NASA's EEO Program as a "near-total failure," with minority representation in the workforce accounting for the lowest of all federal government agencies.[27]

On October 11, Fletcher fired Harris, transferred Hogan to another department, and told Lynn he could stay, provided he could get along with Dudley McConnell, an African American physicist who had moved over to the EEO Office from NASA's Scientific and Technical Information Office.[28] In an unprecedented move, Fletcher also sent a four-page memo to all NASA employees explaining the reasons for Harris's termination. The cause for Harris's release, he claimed, was not the report, but his belief that although she was "an excellent advocate for the cause of minorities," she did not have the "necessary administrative and management skills for the position."[29]

Three consecutive congressional hearings followed Harris's dismissal, beginning in January 1974. Harris and her coauthors testified that despite some "beautiful rhetoric" by NASA's top management, there were concerns that it was just talk and that the necessary changes would not be implemented. NASA, they claimed, considered the EEO Office to be a nuisance and a contaminant to the system.[30]

In his own testimony, Dudley McConnell countered that two factors made the EEO Office job particularly challenging for NASA. First, since 1967, the agency's total employment decreased from 34,000 to less than 25,000 by the end of fiscal year 1974. The total number of hires, therefore, had been far smaller than it would have been during a period of growth. Dramatic increases in the representation of any group would be difficult in such circumstances. Second, NASA required a higher percentage of highly specialized scientists, engineers, and professional administrators—areas in which the availability of minorities and women was limited—and a lower percentage of nonprofessionals in comparison with other government agencies. These two factors, McConnell maintained, produced misleading statistics when attempting comparisons of workforce diversity. He did note, however, that 3.5 percent of scientists and engineers in the United States were minorities and that NASA's 3.4 percent gave the agency numbers comparable to the norm.[31]

Behind the scenes, Harris and NASA carried on negotiations that would bring her back to work at NASA and, if all went well, allow the space agency to calm the tempest surrounding its hiring practices.[32] In his August 16, 1974, memorandum, Fletcher stated that Harris would rejoin NASA as the deputy assistant administrator for public affairs. Her responsibilities would include acting as a major point of contact between NASA and state and local governments as well as community groups, including minorities, women, senior citizens, and people with disabilities. She would also be involved in developing initiatives aimed at increasing the quality of engineering and science education at universities with significant or predominantly minority or female enrollment.[33] Harris had been placed in a position to generate interest in the mathematics and sciences among NASA's underrepresented groups.

In 1977, NASA repeated its claim of a short supply and high demand for women and minorities in the engineering profession.[34] Just as it had during the 1975 hearings regarding Harris, NASA cited increasing minority employment in the face of overall decreases in its personnel. The agency's minority workforce increased from 7.4 percent the previous year to 7.9 percent, while minorities consisted of 5.0 percent of the professionals employed. Still, the agency argued that because of the high number of engineers and scientists in its ranks—49 percent of its workforce—a comparison with other federal government agencies was not a realistic means of measuring the agency's most recent accomplishments in equal employment opportunity.[35]

As part of her contract to recruit minority and women astronauts to NASA, actress–turned–space advocate Nichelle Nichols—Lt. Uhura of Star Trek—visited elementary schools and junior and senior high schools, as well as colleges, to promote space studies among minorities. These youth, she believed, needed to be made aware of the new opportunities that space exploration was opening and that it was possible for them to pursue careers that had been presumed to be beyond their capabilities.[36]

"Historically, blacks have not been involved in science and engineering," she explained, "because you don't need a Ph.D. to work in the post office."[37]

Apathy toward the space program continued to run particularly high among minorities, and NASA's all-male, all-white astronaut corps challenged Nichols's ability to promote a program that provided no role models for the "other." "When you don't see yourself," she explained, "you have no way of feeling involved."[38] Following the imperative to reach minorities while they were still young, Nichols encouraged those who were still adolescent, or not yet qualified, to begin to dream of becoming astronauts or of working in the space program and urged them to take the courses that would qualify them in the future.[39]

On January 16, 1978, NASA administrator Robert E. Frosch announced the selection of thirty-five new astronaut candidates, designated Group 8. Included among them was a groundbreaking group of six women—Anna Fisher, Shannon Lucid, Judith Resnik, Sally Ride, Rhea Sheldon, and Kathryn Sullivan—and four minority men. Three African Americans—Maj. Guion S. Bluford Jr., Maj. Frederick D. Gregory, and Ronald E. McNair, one of the fourteen civilians selected—joined one Asian American, Lt. Col. Ellison Shoji Onizuka, as the first male minority astronauts.[40]

Following his historic 1983 flight as the first African American in space,[41] Bluford took part in several public relations trips, being sure to thank the American people for providing him with the opportunity to fly in space. He spoke with schoolchildren in his native Philadelphia, using his experiences as a means for promoting the significance of the space program and the importance of studying math and science. Bluford stated that he felt very privileged to have been a role model for many youngsters, including African Americans, who aspired to be the future scientists, engineers, and astronauts in the United States.[42]

Twenty years after NASA integrated its astronaut corps, the number of African American participants remained relatively low. Only 6 out of the 119 astronauts in 1999 were black. Bluford and other advocates for greater minority inclusion did not, however, blame NASA entirely for the small level of representation. Instead, they faulted a continuing lack of interest among minority youth in math and the sciences, a resulting lack of role models to be emulated, and a dearth of encouragement from parents and teachers to pursue challenging academic paths.[43]

The collective efforts of NASA and black activists have ensured that the gates of opportunity are open. Toward that end, NASA has granted tens of millions of dollars for academic scholarships and fellowships. The agency has continued its extensive outreach campaign of African American scientists and technicians promoting a future in space exploration to black youth. But, as the space agency moves into a new millennium, it has continued to struggle in its attempts to create a supply of minority engineers to meet its demand.

The challenge, however, has shifted away from a lack of opportunity for minority education to apathy on the part of NASA's target group. The space agency has failed to generate sufficient interest among minority students in the necessary academic fields, despite the increased presence of minorities and women role models to emulate. Ongoing educational inequalities still prevent many from considering the possibilities of walking through the gates at NASA, while a persistent hesitancy and lack of faith in the possibilities appear to hamper the desire of many others to even make the attempt. The space agency's vision statement proclaims the goal of reaching "for new heights." NASA's biggest challenge, however, may be in overcoming the historical reluctance of black youth to join the agency in that pursuit.

Notes

1. Lyndon B. Johnson, "Commencement Address at Howard University: 'To Fulfill These Rights,'" June 4, 1965, *American Presidency Project*, http://www.presidency.ucsb.edu (hereafter cited as *APP*).

2. Roger D. Launius, "Managing the Unmanageable: Apollo, Space-Age Management and American Social Problems," *Space Policy* 24 (2008): 158.

3. "Old Thinking Now Outmoded," *Baltimore Afro-American*, November 16, 1957, 4. See also Emory O. Jackson, "The Tip Off," *Atlanta Daily World*, November 8, 1957, 3; "Truman Rates Civil Rights First and the Sputnik Second," *Philadelphia Tribune*, November 16, 1957, 1, 8; "'Mississippiitis' Blamed for U.S. Lag in Science," *Baltimore Afro-American*, November 16, 1957, 1.

4. James L. Hicks, "Secret Weapon," *New York Amsterdam News*, November 9, 1957, 7; Art Sears Jr., "'We Won't Slow Down,' . . . Marshall," *Cleveland Call and Post*, November 9, 1957, 3A.

5. "Memo from Julius A. Thomas to Lester B. Granger re: Project Proposal in Vocational Guidance, 31 January 1957," NUL Files, Library of Congress (hereafter cited as LOC), Part I, Box A52, Folder: Tomorrow's Scientists and Technicians (TST), 1958, January–April.

6. "Memo from Julius A. Thomas, Director of Industrial Relations, to Lester B. Granger and Ann Tanneyhill re: Observations on Local League Reaction to TST Project, 4 March 1958," NUL Files, LOC, Part II, Box E26, Folder: TST; "Memo from James Rorty to Guichard Parris and Draft leaflet, TST, 22 April 1958," NUL Files, LOC, Part II, Box E26, Folder: TST; "Tomorrow's Scientists and Technicians: A Nationwide Youth-Incentive Program, 15 February 1958," NUL Files, LOC, Part I, Box A52, Folder: TST, 1958 Misc.

7. "Tomorrow's Scientists and Technicians, a nationwide youth-incentive program designed to stimulate and motivate Negro youth toward educational achievements, Draft, 12 February 1958," NUL Files, LOC, Part VI, Box A193, Folder: Southern Regional Office, General Office File, 1958 Tomorrow's Scientists and Technicians Program, February–June; "Tomorrow's Scientists and Technicians: A Nationwide Youth-Incentive Program, 15 February 1958," NUL Files, LOC, Part I, Box A52, Folder: TST, 1958 Misc.

8. Draft "Tomorrow's Scientists and Technicians, a nationwide youth-incentive program designed to stimulate and motivate Negro youth toward educational achievements, 12 February 1958," NUL Files, LOC, Part 6, A193, Folder: Southern Regional Office, General Office File, 1958 Tomorrow's Scientists & Technicians Program, February–June; "Tomorrow's Scientists and Technicians Fact Sheet, June 1958," NUL Files, LOC, Part 1, Box G1, Folder: Vocational Services, General Office File, "Appointment with Tomorrow," 1959; "Tomorrow's Scientists and Technicians, Draft, Leaflet," NUL Files, LOC, Part 2, Box E26, Folder: TST.

9. "Tomorrow's Scientists and Technicians Fact Sheet, June 1958," NUL Files, LOC, Part 1, Box G1, Folder: Vocational Services, General Office File, "Appointment with Tomorrow," 1959; "You Can Be There," Script, NUL Files, LOC, Part 1, Box G13, Vocational Services Office, Projects & Programs, TST, Columbus, 1961.

10. "Memo from Otis E. Finley, Jr., Vocational Services, NUL, to Mr. Granger, 15 November 1960," NUL Files, LOC, Part I, Box A64, Folder: Administration Department, Internal Departments File, Industrial Relations Dept. Memos-Reports, 1960 May–December.

11. "Memo from Otis E. Finley, Jr., Vocational Services, NUL, to Mr. Granger, 15 November 1960," NUL Files, LOC, Part I, Box A64, Folder: Administration Department, Internal Departments File, Industrial Relations Dept. Memos-Reports, 1960 May–December; "New Tomorrow's Scientists and Technicians Career Club Seeks to Motivate Cleveland Youth," *Stride* 1 (August 1959): 1, in NUL Files, LOC, Part 1, Box G12, Vocational Services Office, Projects & Programs, TST, Cleveland, 1958–61; "Monthly Report of Activities, June [1960]," NUL Files, LOC, Part 1, Box G12, Vocational Services Office, Projects & Programs, TST, Cleveland, 1958–61.

12. "Minutes Meeting on Vocational Preparation for Teenagers Held at Hall," October 26, 1960, NUL Files, LOC, Part 1, Box G12, Vocational Services Offices, Projects and Programs, TST, Bridgeport, 1960–61.

13. "INFORMATION MEMORANDUM: The Employment Outlook in Relation to Current Economic Trends, 23 May 1957," NUL Files, LOC, Part I, Box A63, Folder: Administration Department, Internal Departments File, Industrial Relations Department

Memos-Reports, 1957, April–June; Eli Ginzberg, *The Negro Potential* (New York: Columbia University Press, 1956), 100–102.

14. John F. Kennedy, "Special Message to the Congress on Urgent National Needs, 25 May 1961," *APP*.

15. Joseph D. Atkinson and Jay M. Shafritz, *The Real Stuff* (New York: Praeger, 1985), 2.

16. "Management Manual: Administrative Regulations and Procedures, No. 6-3-2, NASA Employment Policy Program, 21 September 1959," Box: NACA/NASA Equal Employment Opportunity Programs, 1923–1992 #18897, Folder: LARC EEO 1960s, No. 3, National Aeronautics and Space Administration Headquarters, Historical Records Collection (hereafter cited as NASA HQ, HRC).

17. "Memorandum from James E. Webb to All Program Directors and Staff Officers, Headquarters; All Directors of Field Installations re: Equal Employment Opportunity Executive Order 10925, 11 April 1961," Folder: LARC EEO 1960s, No. 2, Box: NACA/NASA Equal Employment Opportunity Programs, 1923–1992 #18897, NASA HQ, HRC.

18. "Current Status: Negro Employment [1961]," Folder: LARC EEO No. 2, Box: NACA/NASA EEO Programs, 1923–1992, #18897, 1 of 2; "Letter from James E. Webb to Floyd L. Thompson, Director, Langley Research Center, 12 December 1961," Folder: NACA + Women and Minorities 2, Box: NACA/NASA Equal Employment Opportunity Programs, 1923–1992 #18897, NASA HQ, HRC.

19. James Hicks, "Race and Space," *New York Amsterdam News*, April 16, 1961, 12.

20. "Letter from Paul F. Bikle, Director Flight Research Center, Edwards, California, to NASA Administrator James E. Webb, 28 December 1961," Folder 8983, Minority Groups (1961–1993), NASA HQ, HRC.

21. James L. Hicks, "Negro Math Expert Helped Launch US Spaceman: Her Science Paper Is Key to Man in Orbit," *New York Amsterdam News*, May 13, 1961, 1, 12.

22. Alice A. Dunnigan, "Negro Women Technicians Help Chart Astronauts Course," *Chicago Daily Defender*, July 3, 1963, 37.

23. "John F. Kennedy Space Center, Ben Hursey, Personnel Officer," Folder: NASA EEO ca. 1960s, Federal Records Center, Box: NACA/NASA Equal Employment Opportunity Programs, 1923–1992 #18897, NASA HQ, HRC.

24. Steven L. Moss, "NASA and Racial Equality in the South, 1961–1968" (Master's thesis, Texas Tech University, 1997), 63.

25. "Personnel Change Announcement re: appointment of Ruth Bates Harris as director of EEO," August 24, 1971, Folder 8983, Minority Groups (1961–1993), NACA/NASA Equal Employment Opportunity Programs, 1923–1992, #18897, NASA HQ, HRC.

26. Ruth Bates Harris, *Harlem Princess: The Story of Harry Delaney's Daughter* (New York: Vantage Press, 1991), 254.

27. "Letter and Report from Ruth Bates Harris, Deputy Assistant Administrator for Equal Opportunity Programs et al to James C. Fletcher, Administrator," September 20, 1973, NASA HQ, HRC.

28. Harris, *Harlem Princess*, 270.

29. "Memorandum to All NASA Employees from James C. Fletcher," November 2,

1973, Box: Equal Employment Opportunity, EEO, Box 4 of 4, Folder: Equal Opportunity Programs, Office of General 1968–2001, #18507, NASA HQ, HRC.

30. Congress, Senate, Committee on Appropriations, 93d Cong., 2d sess., January 11, 1974, 69, 83, 86.

31. Ibid., 97–98, 135.

32. "Memo from R. Tenney Johnson, General Counsel to Administrator with Attachment of Mrs. Harris's Demands," April 4, 1974, George M. Low Papers, EEO File, Record Nos. 013696–013704, Folder: Background on R. B. Harris Termination, 013701, NASA HQ, HRC.

33. "Memorandum to All NASA Employees from James C. Fletcher, Administrator," August 16, 1974, NACA/NASA Equal Employment Opportunity Programs, 1923–1992, #18897, Box 1 of 2, NASA HQ, HRC.

34. "Letter from E. S. Groo, Associate Administrator for Center Operations, to Robert L. Krieger, Director, Wallops Flight Center, NASA," February 22, 1977, Folder 8983, Minority Groups (1961–1993), Box: Equal Employment Opportunity (EEO), Code U and DE Chron Files, 1964–1974, #18508, Box 1, NASA HQ, HRC.

35. "Memo from Peter H. Chen, Deputy Assistant Administrator for EO Programs, to Dianne Lambert, Office of Legislative Affairs re: Response to Sen. Proxmire's Letter of 17 February 1977," March 4, 1977, Folder: Code U Chron Files, 1977, Box: Equal Employment Opportunity (EEO), Code U and DE Chron Files, 1975–1982, #18505, Box 2, NASA HQ, HRC.

36. W3 Public Relations [Fact Sheet], Profile: Nichelle Nichols, undated [ca. 1977], File 1594, Nichols, Nichelle (Star Trek), Box: NACA/NASA Equal Employment Opportunity Programs, 1923–1992 #18897, Box 1 of 2, NASA HQ, HRC.

37. "Lt. Uhura Finds a New 'Enterprise,'" Sentinel Star, May 4, 1977, File 1594, Nichols, Nichelle (Star Trek), Box: NACA/NASA Equal Employment Opportunity Programs, 1923–1992 #18897, Box 1 of 2, NASA HQ, HRC.

38. Ibid.

39. Judy Stein, "'Star Trek' Actress Is Successful as Recruiter for Space Agency," National Inquirer, October 18, 1977, File 1594, Nichols, Nichelle (Star Trek), Box: NACA/NASA Equal Employment Opportunity Programs, 1923–1992 #18897, Box 1 of 2, NASA HQ, HRC.

40. "NASA News Release: NASA Selects 35 Astronaut Candidates File 1526, Gregory, Frederick D. (NASA Astronaut)," January 16, 1978, Box: NACA/NASA Equal Employment Opportunity Programs, 1923–1992 #18897, Box 1 of 2, NASA HQ, HRC.

41. Three years earlier, on September 18, 1980, the Soviet Union launched Cuban-born Arnaldo Tamayo Méndez—"the first black cosmonaut"—aboard Soyuz 38. See "Soviets Launch World's First Black Cosmonaut," Jet, October 9, 1980, 8.

42. Guion S. Bluford Jr., interviewed by Jennifer Ross-Nazzal, Houston, August 2, 2004, Oral History 2 Transcript, NASA Johnson Space Center Oral History Project, 26–27, 51.

43. Lisa Hoffman, "Few Blacks Found in NASA's Astronaut Corps," Pittsburgh Post-Gazette, February 22, 1999, A9, File 16593, Minority Groups, 1994–, NASA HQ, HRC.

"Petite Engineer Likes Math, Music"

CHRISTINA K. ROBERTS

In the spring of 1962, American industries urgently needed engineers. The National Aeronautics and Space Administration (NASA) was no exception and boasted of conducting "the biggest manhunt in history" for new hires. In fact, NASA recruiters were "even trying to interest girls and women in engineering" and said hiring women was possibly the only way to fill their engineering gap.[1] It might be surprising that NASA acknowledged the need to hire women engineers in the early 1960s, since scholarly research has shown that NASA performed poorly at hiring women during the long civil rights era. The case for this negative view is based on NASA's low recruitment numbers in comparison to other federal agencies.[2] It has also been suggested that the agency's poor recruitment efforts were the product of an entrenched white male corporate culture that resisted hiring women and minorities into the early 1970s.[3] However, NASA's early recruitment message to women scientists and engineers has not been investigated. Previously unexamined public speeches, news releases, and newspaper articles from 1959 to 1968 invited qualified women to obtain professional science and engineering jobs. That message was transmitted to the public in speeches given by top NASA officials to national women's groups, shared in NASA news releases from the Kennedy Space Center (KSC), and published in newspapers around the country.

In recent years, scholarship has explored the experiences of women such as Katherine M. Johnson, an African American "hidden figure" of the early space program. Johnson was a mathematics computer who started work in 1953 at the Langley Research Center's Guidance and Navigation Department.[4] Scholarship of this type is one way to respond to Jacquelyn Dowd Hall's call for historians to use "modes of writing and speaking that emphasize individual agency" and to dramatize "the hidden history of policies and institutions—the publicly sanctioned choices that continually shape and reshape the social landscape." Hall also suggests that scholars should not "settle for simple dichotomies" and instead should "forego easy closure and satisfying upward or downward arcs."[5] An examination of NASA's recruitment message to women is important because it contains professional and social cues about their prospects for professional science and engineering employment. These cues reflect the complexity of women's lived experience as they entered the agency's male-dominated workforce and thereby changed the social landscape around them.

To begin, officials in the newly created agency transmitted a welcoming message in speeches given by various NASA administrators at women's events. In January 1959, Wernher von Braun gave a speech titled "Women's Role in a Changing World" at the Women's Forum on National Security in Washington, DC.[6] In his speech, von Braun explained that science and technology helped protect the nation during the Cold War. He argued that mothers were especially qualified to guide their children into educational fields that would "provide a fundamental understanding of mathematics and the physical sciences." He told the attendees that "young women caught up in the enthusiasm" for "rocketry and the exciting prospects of outer space exploration" had "many opportunities for participation." Von Braun's message to women about their science and engineering job prospects was positive.

In reality, women were already working in the space industry. Von Braun explained, "One out of every four persons employed in the Army's missile and space projects is a woman." Indeed, they worked as "mathematicians, physicists, chemists, accountants, laboratory assistants." He said that some women plotted "trajectories for satellites and deep space probes." Others calculated flight trajectories for ballistic missiles. Von Braun spoke of numerous opportunities for women at NASA and said that the enormous budget of "the burgeoning missile industry" would "by no means exhaust the possibilities" of employment open to their "sex." According to von Braun, in 1959 women already aided the country as NASA scientists and engineers, and they had many opportunities for meaningful work in the aerospace industry.

Wernher von Braun's 1959 speech coincides with another prevalent message in the 1950s: national security would be stronger if women were educated for science and engineering jobs. The historian Margaret W. Rossiter discusses the vocational push toward science and engineering in the 1950s. It was "the government's official policy to encourage women to enter scientific and technical fields and to urge employers, including the federal government itself, to hire and utilize them fully." However, the vocational push also prompted media discourse about "potentially disastrous consequences that science inflicts on femininity."[7] Women who considered working in the field might have been discouraged by the conflicting messages of the government and the media.

In June 1963, D. Brainerd Holmes, director of Manned Space Flight, gave the commencement speech at Wells College, a private women's college in New York.[8] Holmes said, "There are about 150 women scientists and engineers" on NASA's staff, and "many more are employed by the contractors throughout the country who perform the bulk of our work." He encouraged Wells College graduates to consider working in aerospace because there were "many interesting career opportunities for those of you who complete the necessary preparation." He observed that during the rapidly expanding Space Age, "both men and women—are entering their careers at a time of enormous challenge and change." Holmes characterized their future work as a career, which indicated that women had long-term prospects for professional science and engineering employment at NASA.

In May 1966, Mac C. Adams, associate administrator for advanced research and technology at NASA, addressed the American Institute of Aeronautics and Astronautics.[9] He concluded by recognizing the Langley Research Center's "significant contributions to the progress of aviation and to space." He said the two threads that ran through Langley's distinguished record were the great benefits that research brought and Langley's concern for maintaining the proper research environment. According to Adams, the proper environment was comprised of "first of all, a research team of men and women enthusiastic about their role." Adams's speech indicated that women scientists and engineers were already a respected presence at the Langley Research Center.

In fact, women scientists and engineers regularly received professional acknowledgment in the employee newsletter, *Langley Researcher*. The newsletter published the achievements of women scientists and engineers who had worked at Langley for nearly two decades by the mid-1960s and made announcements about advanced science degrees earned by women employees. The newsletter also provided continuous coverage of the development and implementation of NASA's Program for Equal Employment and Opportunities for Women during the second half of the 1960s. While this news was intended for space industry insiders, employees may have amplified NASA's welcoming message by sharing the news of their achievements and job opportunities with women outside the NASA community.[10]

The welcoming message to women from NASA's top administrators continued in the late 1960s. In July 1968, Thomas Paine, deputy administrator of NASA, gave a speech to the National Federation of Business and Professional Women's Club in Minneapolis.[11] He told the assembly that "women have played a prominent role in the space program in the United States." He said, "within the NASA organization, we have many, many women scientists and executives in prominent positions, people like Mrs. Nancy Roman, who heads up our astronomical program." Paine suggested that there would soon be "even more opportunities for women to make contributions to the space program" and that it was NASA's "sincere hope that more women will be attracted to" working at NASA.

The News about Women Scientists and Engineers Who Worked at NASA

Paine's mention of Nancy Roman is significant because she was a nationally recognized, award-winning, and highly esteemed scientist. News reports about Roman's research and her role as the NASA chief astronomer were regular features in American newspapers. An early mention of Nancy Roman is in the *Alpine (Texas) Avalanche* in 1951.[12] Readers learned that after a month's visit, Roman left the McDonald Observatory in Fort Davis, Texas, for the Yerkes Observatory in Williams Bay, Wisconsin. Roman graduated in 1949 with a doctoral degree from the University of Chicago and worked at the Naval Research Laboratory from 1955 to 1959. She moved on from there to set up NASA's astronomy department soon after the space agency was established.[13]

Before Roman left the Naval Research Laboratory, she was featured in a news article that described women's participation in space science research. In "Cradle Rocking Hand Helping to Put Man Aloft on the Moon," journalist Patricia Wiggins explained that "women scientists and researchers" involved in the space program were "astronomers, biologists, physicists, mathematicians, chemists, draftsmen, physiologists and audiologists."[14] Wiggins said they performed important jobs like "charting 'road maps' to the moon; designing electronic brains for missiles and rockets; developing foods for space-chamber dining; computing orbits for manmade satellites; analyzing cosmic rays and drafting missile parts." She explained that Roman worked "with an 84-foot radio telescope, the largest of its kind in the world," to supply data that differentiates rocket trajectories from natural phenomena in space. She also said Roman measured the temperature of the Moon's surface, a detail Roman found "fairly important if we ever get there."

By late 1960, Nancy Roman was the standard for the claim that NASA did not practice "sex bigotry or prejudice" since she was the new chief of the astronomical and astrophysical program. The issue of "sex bigotry and prejudice" was newsworthy at that time because of the controversy over whether NASA would train women astronauts.[15] The journalist Joseph L. Myler advised that women had other opportunities at NASA besides becoming astronauts.[16] He explained that NASA filled the top astronomy job without consideration of sex. Myler listed the qualifications for Roman's astronomy position: "Must have doctor's degree in astronomy . . . much astronomical experience . . . high intelligence and imagination . . . great common sense and large executive capacity; must have won the respect of astronomers the world over." Nancy Roman fit the criteria, "not because she happened to be a woman, but because she was a fine scientist and organizer." There could be no claim of sex bigotry or prejudice with respect to Roman's employment at NASA.

Roman's scientific expertise was displayed in several articles published in the early 1960s. In 1961, Roman gave a speech to the International Astronomical Union about launching and studying a simulated melting comet.[17] In 1962, she won one of the first Federal Women's Awards, which recognized superior government service by women. She was also recognized for her work on the first Orbital Solar Observatory (OSO) satellite mission.[18] In 1963, she explained what it took for women scientists to work at NASA in the article "Space Program Open to Women."[19] Readers learned that "NASA's requirements for women scientists are exactly the same as for men." Roman said, "It is strictly a matter of merit—education and experience—without regard to sex." Other articles describe the knowledge that the country gained from her research. In 1963, Roman's work with the solar observatory showed that it might be possible to predict solar flares. She also suggested that it was possible to launch an orbiting telescope.[20] Roman later shepherded the project that became the Hubble Telescope mission.[21] Roman attested many times that she found personal fulfillment in her scientific work and that other women scientists and engineers could expect to do the same.

In 1964, the *Hutchinson (Kansas) News* devoted a full page of articles to Roman and seven other women scientists who worked at NASA facilities such as the Jet Propulsion Laboratory, the Langley Research Center, the Goddard Space Flight Center, the Ames Research Center, and the Aerospace Medical Laboratory at Cape Canaveral.[22] In one article, Roman was called a "woman doer," President Johnson's term for "a woman of intelligence and achievement." Another article described Jocelyn Gill as someone who taught, briefed, and debriefed the Mercury astronauts but was also "NASA's In-Flight Sciences chief" who had "purview over the experiments conducted in manned space vehicles."

Yet another article described the work of Melba Roy, an African American woman pictured next to an IBM computer. The journalist said that Roy headed "a team of mathematicians" that formed the computer programming division at the Goddard Space Flight Center, and that beginning in 1959 "she built the team of mathematicians virtually from the ground up." Roy had "nine computer programmers" with "no experience at all," yet she still created a "division consisting of 12 experts" who produced "incredibly accurate maps of orbiting satellites."

The same article introduced Marcia Neugebauer, a JPL employee who "helped develop a solar wind spectrometer carried by Mariner II." Neugebauer worked on the "Ranger series of spacecraft which succeeded in photographing the Moon's surface." There was also "Mrs. Barbara Lunde, a 27-year-old aerospace engineer . . . responsible for the invention of celestial sensors which help orient, or determine the attitude, of satellites in space." Next was Eleanor Crockett Pressley, who oversaw "the work of 15 engineers" and managed "a budget of about $3 million a year" for "NASA's sounding rocket program." Like Nancy Roman, Pressley received the Federal Woman's Award, which was "presented annually to outstanding career women in the Federal government service." These scientists and engineers were all highly educated, professional women who had important roles in the space program. They were portrayed as role models of what women scientists and engineers could achieve at NASA centers around the country.

None of the news articles cited up to this point have been connected to specific NASA news releases. Perhaps the stories originated from NASA's public affairs offices, but further research in NASA archives is needed to test this idea. Journalists may have performed their own investigations. Some articles were written by women journalists; some were located in women's sections of local newspapers. In any case, the administrators' speeches and the newspaper articles about Nancy Roman and others state that women scientists and engineers could expect to find work at NASA from the late 1950s forward. While many journalists noted that women were already doing science and engineering work at NASA centers around the country, in 1960 one journalist noted that the KSC had not yet employed women as other than secretaries.[23] This could explain why the KSC was eager to tell the public about its newly hired women scientists and engineers by the mid-1960s.

News Reporters and Journalists Use NASA News Releases

The KSC distributed thousands of news releases during its first decade of operation. Within that body of information was NASA's welcoming message about women's professional science and engineering employment. The connection between NASA and the newspaper-reading public was mediated by journalists and reporters who used NASA's news releases to write public interest stories about such things as the controversial topic of women working in science and engineering. This section describes NASA's news releases and their subsequent appearance in newspaper articles published around the country.

In February 1965, the KSC's public information office distributed a news release about "petite Jeanette Denny of the Planning and Resource Office" who was "the only woman civil engineer at the NASA Kennedy Space Center." Denny also earned the "first Civil Engineering degree ever awarded at the University of Tennessee." She was one of six women and 700 men enrolled in engineering at that university. Denny was "crowned University Engineering Queen for four successive years" and named "the Engineering Dream Girl" during her studies. The author said about Denny, "She doesn't consider her sex a hindrance in what is predominantly a man's field."[24]

There is evidence of a strict merit-based hiring policy at the KSC in a news release distributed on February 25, 1965.[25] Equal Employment Opportunity coordinator Gene Balstad referenced the policy and said, "We hire new engineers strictly on their qualifications, sex has nothing to do with it." He then described the work of three women employed at the KSC. First was Sally Gruben, an aerospace technologist in the Data Acquisition and Systems Analysis Division. She was a mathematics graduate from the University of Minnesota and was a computer programmer. Next was Coralee Whisenant, who worked in the KSC's Advanced Studies Office and earned a degree in mechanical engineering from the University of Arkansas. Finally, Balstad described Janie Callahan, who earned a master's degree from Texas Christian University. Callahan was an aerospace technologist involved with orbit and trajectory studies. Balstad did not reference the engineers' physical or feminine features; he emphasized their qualifications.

Many of the scientists and engineers mentioned, such as Coralee Whisenant, Cherie Lee, Janie Callahan, Sally Gruben, Jeanette Denny, and Jeanne Johnson, were discussed in the article "A Man-Sized Job Done by Space-Base Women." The article was published around the country in 1965 but was compiled from information found in a few different news releases distributed between May 1964 and February 1965.[26] "A Man-Sized Job Done by Space-Base Women" was published in Racine, Wisconsin; "Women" was changed to "Ladies" in State Center, Iowa, in Hamilton Ohio, and in Bay Town, Texas; and to "Gals" in San Francisco. The newspaper articles were originally printed in May, June, September, and November 1965.[27]

Mary Driver, the personnel staffing specialist at the KSC, had the same attitude as the Equal Employment Opportunity coordinator, Gene Balstad. Driver did not hire based on feminine traits but on professional merit. This is clear from her comments in a news release distributed in April 1966 titled "Women Have a Place in the Space Program."[28] Driver had worked at NASA since 1961 and was responsible for making hiring decisions about recruits who possessed highly technical skills. The news release made it clear that Driver had a strong conviction that women's skills were "basically equal to those of men." In fact, Driver had just placed a "female employee in a highly responsible contract administration position" and had recruited "a female engineer" doing "microscopy work in reliability investigations on spacecraft systems."

Later in 1966, Driver provided a progress update on hiring women at the KSC in a news release titled "Opportunities for Women Engineers Are Growing."[29] She said, "It's been a good year for hiring girls. . . . In fact it's been our best. Hiring six lady engineers in a single year is a record here." The news release noted that "a total of 75 women now occupy professional, administrative or technical positions, and all of them—including a total of 15 engineers—work at tasks usually handled by the male sex." Driver concluded, "Here at KSC we're finding that in our job positions, sex makes no difference—women are just as employable as men."

A widely published article was written about Janie Callahan, based on the news release sent by the KSC in December 1966 titled "Space Has Opened Doors for Qualified Women."[30] A commonly used newspaper article title was "Technologist Traces Saturn's Path." At least a couple of newspapers informed readers that Callahan was a "Lady Technologist." The articles were printed in Richardson, Texas; Cedar Rapids, Iowa; Sandusky, Ohio; Tipton, Indiana; and Fitchburg, Massachusetts, during May, June, and July 1968. The articles used information directly from the news release that the KSC distributed at the end of 1966. The time between distribution and publication shows that journalists sometimes composed articles well after NASA transmitted the information.

The title of this chapter comes from a KSC news release titled "Petite Engineer Likes Math, Music," which was distributed on July 18, 1968.[31] The news release was about an engineer named Cherie Lee, a 1967 math graduate of the Florida Institute of Technology. She worked in the Automation and Programming Office, Launch Vehicle Operations, at the KSC. Another news release was about Jeanne H. Johnson, an aerospace technologist. Johnson was introduced as a "Lady Engineer at Space Center" who "Hopes for Another Degree." The author of the news release wrote, after describing the "lady engineer's" views on education and her work, "Jeanne definitely does not consider her sex a handicap."[32]

Finally, there were other published newspaper articles that focused solely on Jeanne H. Johnson. The KSC distributed a news release titled "Lady Engineer at Space Center Hopes for Another Degree" in January 1968.[33] Johnson's story was published in newspapers all during the same year that the news releases were distributed. The newspaper articles were titled "A Girl's Career in Space," "The Lady's a Space Engineer," and "Jeanne Finds Time for Double Role in Space." They were published in Clearfield, Pennsylvania, and Richardson, Texas.[34] Johnson was described as an "auburn-haired, bright-eyed girl" who "just turned 24 years old." She "enjoyed her multiple roles as housewife, student, and working engineer during her two years of marriage" and loved "to cook and sew." Interestingly, Johnson was not the only woman described in terms of her feminine qualities or activities.

The documentation shows that the characterization of female engineers as petite or pretty ladies with feminine qualities started to trend in the mid-1960s. Most of the NASA news releases include descriptions of feminine traits, interests, and activities for each of the women scientists and engineers described, while only a couple of them did not. The feminine details were either observed by the news release author or provided directly by the women themselves. Within the newspaper articles, journalists sometimes added their own feminine qualifiers to the discussion about the women's advanced science and engineering skills.[35]

The historian Amy Bix provides insight into this phenomenon in her work about the experiences of college women in twentieth-century engineering programs.[36] Bix also discusses the Society of Women Engineers (SWE) emphasis on femininity. She writes, "at a time when many Americans perceived female engineers as odd, manlike creatures, SWE representatives took pains to offer a presentable feminine image, emphasizing that many of them were married and had children." Bix details how the SWE actively tried to change the public perception that women who entered science and engineering degree programs in college were odd or unwomanly. In the 1960s, the SWE increased its efforts to provide "both encouragement and information about women's place in the field. Among other things, its pamphlets documented that female engineers were not abnormal and were not a bunch of old maids." Whether the scientists and engineers at NASA were members of the SWE or not, the positive tone and the feminine details of the news releases and newspaper articles suggest they would have at least understood the SWE's efforts.

Exploring previously unexamined sources has shown that NASA distributed a positive message about its desire to employ women scientists and engineers from its early days forward. That message was transmitted in official speeches beginning in 1959 and in news releases from the KSC beginning in 1964. The stories that NASA disseminated in news releases were used by the newspaper press and published all over the country.[37] NASA also portrayed a culture of respect for women scientists and engineers in the Langley Research Center newsletter.

These sources deepen our understanding of women's prospects for professional science and engineering employment at NASA during the long civil rights era because they profess that a positive experience awaited women. This portrayal may conflict with many other women's struggles to break into male-dominated professions and reflects the complexity of professional women's experiences. The sources also indicate that women possessed and needed professional merit and femininity as NASA employees; understanding these social and professional cues may expand our interpretation of their experiences.[38] Certainly, the struggles and challenges of entering a traditionally male workplace should not be diminished, even though women were simultaneously recruited, publicly encouraged, and welcomed to work at NASA.

It should be noted that none of the documents cited advertised opportunities for women astronauts or specifically appealed to women of color to join the science and engineering teams at NASA.[39] Therefore, the public message that women were welcome, while evident, seems to have been directed at white female science and engineering college graduates who sought a professional career in the space industry. Also, the news releases cited were not generated until 1964, the same year of the Civil Rights Act, Title VII, and the establishment of the Equal Employment Opportunity Commission.

In conclusion, previous scholarship has demonstrated that NASA had significantly poor hiring practices regarding women scientists and engineers. However, as the sources in this essay also show, beginning in the late 1950s the agency publicly recruited women scientists and engineers in speeches, news releases, and newspaper articles. The American newspaper-reading public learned of the activities of women scientists and engineers throughout the 1960s and knew that women held important professional and administrative jobs at the agency. The welcoming message in public discourse needs more critical examination in the conversation about women's experiences at NASA during the long civil rights era.

Notes

1. "Biggest Manhunt: Race for Engineers Grows Hotter," *Cedar Rapids (IA) Gazette*, May 6, 1962.

2. J. Edward Kellough, "Federal Agencies and Affirmative Action for Blacks and Women," *Social Science Quarterly* 71, no. 1 (March 1990), Jstor, http://www.jstor.org/stable/42863588. Kellough compared the rates of hiring improvement in several federal agencies and found that NASA had significantly increased hiring women and people of color, but this was because NASA was very far behind, with much room for improvement after the 1971 application of goals and timetables for affirmative action.

3. Kim McQuaid, "'Racism, Sexism, and Space Ventures': Civil Rights at NASA in the Nixon Era and Beyond," in *The Societal Impact of Space Flight*, ed. Steven J. Dick and Roger D. Launius (Washington, DC: NASA Office of External Relations History Division, 2007), 421–449. McQuaid notes that "until September 1971, NASA had no systematic civil rights element in its employment program" (425). Also, Nancy Schwartz discussed the scarcity of women scientists and engineers at the Kennedy Space Center during the 1960s in "'A Man's World?': A Study of Female Workers at NASA'S Kennedy Space Center" (Master's thesis, University of Central Florida, 2002). She reviewed the center's employee newsletter, *Spaceport News*, which published, among other topics, special-interest stories about rare women science and engineering employees.

4. See https://www.nasa.gov/feature/katherine-johnson-the-girl-who-loved-to-count. See also the recent book by Margot Lee Shetterly, *Hidden Figures: The American Dream and the Untold Story of the Black Women Who Helped Win the Space Race* (New York: William Morrow, 2016).

5. Jacquelyn Dowd Hall, "The Long Civil Rights Movement and the Political Uses of the Past," *Journal of American History* (March 2005): 1233–1263.

6. All of the speeches referenced in this chapter are archived online. They were gathered from https://historydms.hq.nasa.gov/content/speeches-key-officials.

7. Margaret W. Rossiter, *Women Scientists in America: Before Affirmative Action 1940–1972* (Baltimore: Johns Hopkins University Press, 1995), 54. Rossiter also argues that this was mostly a rhetorical message because the government did not provide federal enforcement or incentives to comply with the official position. By 1957, the total number of women enrolled in science and engineering was still below 1 percent (56).

8. D. Brainerd Holmes, Commencement Exercises Wells College, (1963).

9. Mac C. Adams, "NASA Research and Technology for Future Missions" (1966).

10. See articles in the newsletter *Langley Researcher* from 1963 through 1965, https://crgis.ndc.nasa.gov/historic/People#Bulletins.2C_Scoops.2C_and_Researchers.

11. Thomas Paine, National Federation of Business and Professional Women's Club (1968).

12. All of the news articles referenced in this chapter are digitized and archived online in the database at NewspaperARCHIVE.com. "Fort Davis News," *Alpine (TX) Avalanche*, September 28, 1951.

13. A short biography is located at https://women.nasa.gov/nancy-grace-roman-2/; see also https://solarsystem.nasa.gov/people/romann.

14. Patricia Wiggins, "Cradle Rocking Hand Helping to Put Man Aloft to the Moon," *Fairborn (OH) Daily Herald*, January 27, 1959.

15. Joseph L. Meyer, "No Girl Astronauts Are Included in Plans of Space Agency," *Lubbock (TX) Avalanche Journal*, September 29, 1960.

16. Myler won an award in 1969 from UPI for his "Significant Contributions to Public Understanding of Atomic Energy, Atomic Industrial Forum, Inc.," https://100years.upi.com/history_awards.html.

17. Alton Blakeslee, "May Test Theory Comets Have Core Like Snow or Ice," *Terre Haute (IN) Star*, August 25, 1961.

18. "She Launches Satellites," *Humboldt Times* (Eureka, CA), July 27, 1962.

19. "Space Program Open to Women," *Altoona (PA) Mirror*, May 14, 1963.

20. "Sun Flare Eruptions Might Be Predictable," *Albuquerque (NM) Journal*, August 15, 1963.

21. See recorded interview at https://svs.gsfc.nasa.gov/12634.

22. "Women Find Challenging Place in the Sun," *Hutchinson (KS) News*, November 29, 1964. This page is located in the Family Living and Features section.

23. See, for example, "Many Women Employed at NASA Centers," *Laredo (TX) Times*, December 25, 1960.

24. The KSC's news release archive is available online. All of the referenced news releases were gathered from https://www.nasa.gov/centers/kennedy/news/releases/1960/index.html. KSC-34-65, no title, February 25, 1965. There were also comprehensive, individual stories in news releases distributed in 1964 about Coralee Whisenant and in 1966 about Janie Callahan. KSC-62-64, no title (Whisenant), May 6, 1964; KSC-275-66, "Space Has Opened Doors for Qualified Women" (Callahan), December 11, 1966.

25. KSC-37-65, no title, February 25, 1965.

26. KSC-62-64, no title, May 6, 1964; KSC-37-65, no title, February 25, 1965; KSC-34-65, no title, February 25, 1965.

27. "A Man-Sized Job Done by Space-Base Women," *Racine (WI) Sunday Bulletin*, May 23, 1965; "A Man-Size Job Done by Space Base Ladies," *State Center (IA) Enterprise*, May 13, 1965; "A Man-Size Job Done by Space Base Ladies," *Baytown (TX) Sun*, June 23, 1965; Frank Macomber, "A Man-Size Job Done by Space Base Women," *Hamilton (OH) Daily News Journal*, November 4, 1965; "Man-Sized Jobs for Space Age Gals," *Pacific Stars and Stripes* (San Francisco), September 29, 1965.

28. KSC-93-66, "Mary Driver: 'Women Have a Place in the Space Program,'" April 13, 1966.

29. KSC-369-68, "Opportunities for Women Engineers Are Growing," August 1, 1966.

30. KSC-275-66, "Space Has Opened Doors for Qualified Women" December 21, 1966; "Lady Technologist Traces Saturn's Path," *Richardson (TX) Daily News*, July 11, 1968; "Technologist Traces Saturn's Path," *Cedar Rapids (IA) Gazette*, May 12, 1968; "Technologist Traces Saturn's Path," *Sandusky (OH) Register*, June 1, 1968; "A Lady Traces Saturn," *Fitchburg (MA) Sentinel*, May 10, 1968.

31. KSC-344-68, "Petite Engineer Likes Math, Music," July 18, 1968.

32. KSC-14-68, "Lady Engineer at Space Center Hopes for Another Degree," January 17, 1968.

33. Ibid.

34. "A Girl's Career in Space," Progress Youth Page, *Progress* (Clearfield, PA), March 26, 1968; "The Lady's a Space Engineer," *Cedar Rapids (IA) Gazette*, April 1, 1968; "Jeanne Finds Time for a Double Role in Space," *Richardson (TX) Daily News*, May 25, 1968.

35. Nancy Schwartz also writes that the *Spaceport News* (employee newsletter) stories characterized women as feminine and pretty, which she interpreted as a societal trait and a typical feature of internal, 1960s KSC work culture. Comparisons to other NASA's employee newsletters at other NASA centers may help explain if this was a KSC phenomenon or was more widely present in the NASA culture. Schwartz, "A Man's World?" The *Langley Researcher* (1963–1965) did not use these characterizations, however.

36. Amy Sue Bix, *Girls Coming to Tech: A History of American Engineering Education for Women* (Cambridge, MA: MIT Press, 2014), 127–128.

37. In fact, all but one of the news releases described here were published. A search returned zero published stories about Mary Driver in the digital archive at newspaperarchive.com and ProQuest Historical Newspapers' database for the *New York Times*. The absence raises questions: Is it possible that journalists were not interested in NASA's message of opportunity if the reassurances about femininity weren't included? Were journalists conscious of the dual message about merit and femininity? Were the details about femininity important to recruits or particularly helpful to the hiring process at NASA during the civil rights era? These questions may need further examination in order to understand what messaging women consumed as they considered entering the ranks of professional NASA employment.

38. Rossiter, *Women Scientists*. Rossiter describes the 1950s media's message about femininity and intelligence, which seems similarly dissonant. The applicable point is that the media addressed and then denied any such conflict between femininity and intelligence. However, the media also portrayed Soviet women scientists and engineers as unattractive, unfeminine loners (ibid., 64–66), which could have disenchanted some American women who considered obtaining science and engineering degrees in the 1950s.

39. See the scholarship published on these topics, such as Margaret A. Weitekamp, *Right Stuff, Wrong Sex: America's First Women in Space Program* (Baltimore: Johns Hopkins University Press, 2006); Nathalia Holt, *Rise of the Rocket Girls: The Women Who Propelled Us, from Missiles to the Moon to Mars* (New York: Little, Brown, 2016); Shetterly, *Hidden Figures*.

Conclusion

Where Do We Go from Here? Ensuring the Past and Future History of Space

JONATHAN COOPERSMITH

Three major challenges of doing history are finding, preserving, and presenting material, challenges particularly acute for subjects traditionally not collected by government archives, such as minority movements. The story of being one of the first black engineers at the National Aeronautics and Space Administration (NASA) died untold with Frank Williams to the anguish of his daughter and Richard Paul and Steven Moss, who could not successfully interview the terminally ill Williams for their *We Could Not Fail: The First African Americans in the Space Program*.[1]

To prevent such future losses, this chapter explores how historians, archivists, and other stakeholders can encourage the collection, preservation, and accessibility of the widest possible range of appropriate records and histories, especially for historically underrepresented and underresearched areas and people in space exploration and exploitation. The goals are fourfold:

- Promote the creation, preservation, curation, and access of the histories and activities of current and retired minorities involved in space;
- Understand how the changing worlds of oral histories and paper-based and electronic records are changing how historians and archivists operate, including the increasing ease of facilitating history from below;

· Encourage contemporary space actors (including institutions like the Prairie View A&M University Center for Radiation Engineering and Science for Space Exploration and the National Society of Black Engineers) to similarly document and preserve their ongoing history; and

· Ensure access and availability of these records to the public as well as academics.

This chapter begins with an overview of the state of archival collecting of African American history in order to frame minority space history in the contexts of digital archiving, minority archiving, and space archives. It concludes with a set of recommendations for further action.

The State of African American Archives

Histories are only as good as their sources and resources. What is impressive, if you talk to space researchers and read the acknowledgments and forewords in their books, is the debt that they—and we—owe to archivists, NASA history offices, and funding agencies. *We Could Not Fail* succeeded because of a National Science Foundation (NSF) grant and a National Air and Space Museum (NASM) Verveille Fellowship. The NASM especially deserves kudos for its willingness to think outside the box in offering a nonacademic, Richard Paul, that fellowship. Steven Moss was able to research and write his 1997 master's thesis, "NASA and Racial Equality in the South, 1961–1968," the basis for *We Could Not Fail*, thanks to grants from the JFK and LBJ presidential libraries.[2] A Sloan Foundation award gave Margot Lee Shatterly the time to write *Hidden Figures*. The Virginia Foundation for the Humanities and the Hampton Roads Chapter of the Association for the Study of African American Life and History are supporting her Human Computer Project.[3]

Rarely do historians think proactively about the state of archives. We accept that the National Archives was, is, and will be chronically underfunded. The 2016 Future of the African American Past Conference had a session titled "History, Preservation, and Public Reckoning in Museums," but it did not discuss the larger challenges and opportunities for creating, preserving, and promoting African American archives.[4]

Compared with government archives, African American archives have historically faced additional burdens, including more acute lack of resources and support in a wide range of areas.[5] In recent decades, however, this gloomy picture has brightened due to the creation of new institutions, organizations, and trained individuals, such as the 1987 establishment of the Archivists and Archives of Color Section of the Society of American Archivists.[6]

New electronic and particularly digital technologies have expanded the worlds of archives. The growing academic world of digital humanities embeds those technologies in an increasingly dense theoretical framework to contextualize, interpret, analyze, and share the past.

The intersection of archives and digital humanities is incredibly exciting. Scores of archives, coalitions, universities, nonprofits, businesses, and other organizations are experimenting with every aspect of the historical process, from collecting data to transcribing documents and creating displays online. While big projects have multimillion-dollar price tags, DIY oral historians can start work with their smartphone and a downloaded list of basic questions.

It is worth concentrating on these digital frontiers because they offer the greatest possibilities for the future. That the digital world builds on and augments—not replaces—the world of physical archives must not be forgotten.

The Digital Public Library of America, evolving since 2013 with the "ambitious goal of bringing together the riches of America's libraries, archives, museums, and cultural heritage sites, and making them freely available to students, teachers, researchers, and the general public," is, like Umerka Search, trying to "desilo" physically separate materials by making them digitally accessible.[7] The results can appear impressive, and filters help reduce the data overload, but for space history, at least, this is an effort low on the learning curve.[8] Its more than thirty Primary Source Sets for K–12 education do not, alas, contain anything on the space program, but that might become a goal for its Education Advisory Committee.[9]

Digital African American materials are growing in number, accessibility, and experimentation.[10] Blackpast.org lists sixty-two African American digital archives.[11] Only "The Faces of Science: African Americans in the Sciences" is devoted to science and inventors (no engineers) and has not been updated since 2007.[12] The University of Minnesota Libraries' Archives and Special Collections has developed Umbra Search, which links over 500,000 African American digital items from over 1,000 libraries and archives.[13] The subscription-only Black Studies Center offers the

Schomburg Studies on the Black Experience, the International Index to Black Periodicals, and other material.[14]

Space archives face major challenges. Hundreds of new state and non-state institutions are active in space exploration and exploitation, but many of these institutions are unfamiliar with the importance of records retention and archiving for historical purposes. Concepts of space history are expanding and incorporating the perspectives of broader communities and the evolution of its many participants. New information technologies continue to expand and transform what can be collected, raising preservation issues never before faced. These new information technologies provide an opportunity to democratize the creation of and access to archives, but only if they are designed for long-term preservation and accessibility. This is a critical time for both archival records and oral histories as institutional and personal records are lost or destroyed and participants in the early years of the space program die.

Donald Rumsfeld, while secretary of defense, was unjustly pilloried for ruminating about known knowns, known unknowns, and unknown unknowns.[15] He actually made a great deal of sense, as demonstrated by viewing the state of archiving African American participation in space by those categories.

Known knowns: Federal, state, local, private, and digital archives offer a wealth of data, especially after the 1964 Civil Rights Act increased government programs focused on African Americans. A challenge here is finding what is there. Digitizing newspapers has transformed searching the African American press.[16]

Known unknowns: These are recent, current, and future sources that we should be collecting. These areas include not just gaps in NASA (and other government agencies) archiving but also the growing worlds of private and nonprofit groups in space. The challenge here is convincing individuals and organizations to dedicate the resources to collect and preserve their history.

Unknown unknowns: These are materials of a rapidly disappearing past, unveiled and promoted in *Hidden Figures* and *We Could Not Fail*. The people may be disappearing, but there are, as those books demonstrated, materials out there. Finding them and figuring out their significance are the primary challenges.

Their acknowledgments show the extraordinary lengths researchers like Richard Paul and Steven Moss of *We Could Not Fail* and Margot

Lee Shetterly of *Hidden Figures* went to find their subjects, capturing the voices now silent forever. Sometimes they waited too long. Paul initially started his research from a 1958 *Ebony* article, "Negroes Who Help Conquer Space: Over 1,000 Negroes Are in Satellite, Missile Field."[17] He tracked down thirty-five of the forty-five men profiled. All were either dead or mentally incapable of cooperating.

Paul described his research as traditional journalistic shoe leather, where "I would find a thread and I would pull and pull and at the end I would find Clyde Foster."[18] As a historian, I thrill when I uncover a key document. When researching Russian electrification in the British Library, one of my biggest pleasures was slitting open the pages of *Elektrichestvo*, the first Russian electrical engineering journal, becoming the first person to read them since they were printed over a century ago. Today digitization and the Internet make *Elektrichestvo* and other information available at my laptop—but only if the materials are processed and made accessible.

These unknown unknown sources are the people we don't even know we are missing. Whose stories are not being found? Who and what are we missing? What other approaches do we need?

Existing hidden collections that have yet to be processed and publicized are one unknown unknown.[19] In the increasingly online world, not having, for example, a local historical society's holdings posted on a website (though not necessarily accessible) by a Google search is the equivalent of local knowledge. Informal networks of knowledge such as alumni and retiree organizations and local networks like church groups are key to reducing these unknowns.

Our goals should be to encourage collecting, preserving, and providing access to the past, contemporary history, and future history. Reducing the barriers to entry is important: not everyone has the expertise or time of Paul or the ability to sit in the British Library. We want to reduce the need (though not eliminate the challenge and excitement) for heroic measures to capture space history. Collecting future history means establishing the foundations now for institutions and individuals to routinely act as vacuum cleaners, scooping up materials to ensure their stories will be captured and preserved.

Collecting and preserving are not enough. We also need to promote and publicize these stories to the rest of the academic community and the outside world. The Internet is the essential foundation on which a growing range of options like the Alliance for Networking Visual Culture offer opportunities to reach audiences.[20] Archives and libraries increasingly post some materials on their websites. Creating a website coordinated with a project or book release is also becoming more common.[21] Class projects can produce websites with primary materials like the 2015 "Race to the Moon" exhibition on the Digital Public Library of America by students in a capstone course at the University of Washington Information School.[22]

Oral histories, an area that has exploded with digital technologies, are one invaluable tool. As the T. Harry Williams Center for Oral History, Louisiana State University, stated,

> Used to its full potential, done carefully and conscientiously, oral history methods let us collect unique information that can be of great value to researchers now and in the future. Done carelessly, without proper preparation and processing, it can result in evidence that is superficial, anecdotal, little better than hearsay and too often inaccessible.[23]

The challenge is to do it well. And with limited resources. The first rule is not to reinvent the wheel. For doing oral history, an excellent place to start is the Oral History in the Digital Age project, http://ohda.matrix. msu.edu/, which offers extensive guidance on best practices for doing oral history digitally, including how to start a project (first question: why do you want to do this?).[24]

The second rule is to work with as well as learn from others. We need to interest people outside as well as inside the humanities to work together, share resources, and provide a forum for presenting results.

Simply creating oral histories is not enough. They must be preserved, housed, and made accessible in archives, which, like everything else, takes resources. Digital archives are far easier to store than physical materials but still demand resources, including challenges of longevity, backup, and access. One goal should be to work with STEM education and museums, including the Association of African American Museums, to house and promote these new collections.[25]

Especially for digital humanities projects and groups involved with underrepresented communities, a sense of social purpose provides a strong underlying motivator. Some projects proclaim they are promoting intersectional "radical institutional partnerships, open data, and the use of technology,"[26] reflecting a bottom-up desire to experiment with the possibilities enabled by the digital humanities. Umbra Search intends to be more than just a search engine. Significantly:

> As much as Umbra Search is a resource for scholars, students, writers, and artists, it is also a call to action for more inclusive collection, description, and digitization efforts and for more partnership to accomplish the work still left to do.[27]

The archival equivalent of "citizen science" is crowdsourcing archival functions because of the lack of resources for archives (this could be viewed as a failure of neoliberalism to fund some sectors of the state sufficiently).[28] The New York Public Library invited people to transcribe its extensive menu collection as a successful experiment starting in 2012.[29] The National Archives and Records Administration (NARA) is a major innovator in crowdsourcing transcribing since its Innovation Center released its Citizen Archivist Dashboard in 2011.[30] Ensuring quality control is a major challenge of such projects.

Could we develop projects to harness the enthusiasm of flash mobs, hackathons, and citizen science to collect, preserve, and promote African American space history? Or are longer-term, more subdued projects more likely to produce long-term benefits?

Some practical measures for expanding the collection of African American space history include:

· Connect undergraduate and graduate students in classes doing oral as well as traditional history to space-related projects.
· Create a network of historians willing to have their students do these projects.
· Focus on Internet-based dissemination as well as traditional papers to reach broader audiences.
· Work with individuals and institutions to capture materials on an ongoing basis.

Preserving the present and recent past is another goal. At the institutional level, one target should be African American space research based in Historically Black Colleges and Universities (HBCUs), like the Prairie View A&M University Center for Radiation Engineering and Science for Space Exploration.[31]

Working with professional societies offers a way to capture a larger slice of space history beyond government activities. A prime opportunity is the Aerospace Systems Conference of the National Society of Black Engineers. Held since 2010, the biennial conference enabled participants not only to hear "the recurring stories of personal and professional challenges overcome by a passion for science, technology, engineering and/or math" but also, in the words of founder Robert L. Howard, the head of the Habitability Design Center at the Johnson Space Center, to be "(1) a generator of advanced aerospace technology of African-American origin and (2) a place for black aerospace executives to make important business deals with each other. To my knowledge, this is something that does not exist anywhere else in the world."[32]

If nothing else, the conference's Celestial Torch awards provide a guide to some of the most prominent black space actors. Can we find the resources and develop the relationships to interview these men and women and convince them and their employers to routinely save key documents?

Other lines of research are professional groups. The National Technical Association, founded in 1925 to "encourage and inspire women, minorities and youth to enter and excel in the fields of math, science and technology," was active at Langley during the National Advisory Committee for Aeronautics (NACA) years.[33] Three other groups that warrant similar attention from archivists and historians are aerospace writers, minority journalists, and academics who covered and are covering the space program. What do they do with their accumulated materials? Should we try to encourage them to donate their papers to archives and to have oral histories done of them?

Resources

Collecting and preserving history takes resources as well as resourcefulness. Oral history can be expensive. The T. Harry Williams Center for Oral History, Louisiana State University, charges $500–$650 for every hour of recording.[34] The History Makers has spent $30 million on 2,700 videotaped oral histories since it was founded by Julieanna Richardson in 2000.[35] Those are gold-standard approaches. But even using volunteers, donated equipment, voice recognition software, free online hosting, and other support takes resources.

One desirable but extremely unlikely option for funding is to copy the 2003 21st Century Nanotechnology Research and Development Act, which mandated a small percent (<1 percent) of the total budget on societal dimensions research into nano-scale science and engineering.[36] Wouldn't a mandated 0.1 percent of the NASA budget fund many projects? At a time when the Johnson Space Center almost lost its history office, this seems a doubtful priority.

More realistic are finding patrons and other enthusiasts, creating connections, linking with STEM programs, crowdsourcing, applying for grants, and publicizing these efforts. For academics, a talk with their college or university fund-raiser may create opportunities for dedicated opportunities to support their work.

Conclusion

The ultimate goal of preserving the past is to inform and enrich the future. How do we know we have succeeded? Perhaps one sign of success is becoming a Lego figure. Earlier this year, Maia Weinstock's "Women of NASA" won the Lego Ideas vote to make Lego figures out of mathematician Katherine Johnson; computer scientist Margaret Hamilton; astronaut, physicist, and educator Sally Ride; astronomer Nancy Grace Roman; and astronaut and physician Mae Jemison.[37] Short of Lego fame, however, what metrics, qualitative and quantitative, should we use to judge how we are doing?

Notes

1. Richard Paul and Steven Moss, *We Could Not Fail: The First African Americans in the Space Program* (Austin: University of Texas Press, 2015), ix–x.

2. Ibid., 283–286; Steven Moss interview, February 10, 2017.

3. Margot Lee Shetterly, *Hidden Figures: The American Dream and the Untold Story of the Black Women Mathematicians Who Helped Win the Space Race* (New York: William Morrow, 2016), 267–271.

4. https://futureafampast.si.edu/sessions/session-7-history-preservation-and-public-reckoning-museums.

5. Including art. Rebecca Hankins and Miguel Juarez, "Art in Special Collections: Latino and African American Fine Art and Photography Collections in Academic Institutions," *Art Documentation: Bulletin of the Art Libraries Society of North America* 29, no. 1 (2010): 31–36, http://hdl.handle.net/1969.1/90986.

6. Society of American Archivists, "Mission and Vision," http://www2.archivists.org/groups/archivists-and-archives-of-color-section/mission-and-vision.

7. https://dp.la/info/about/strategic-plan/.

8. Searching for "NASA" produced 30,087 results from 141 institutions, while "NASA African-American" generated 24 results from 10, including 5 for Guion Bluford—and 4

for Sally Ride: https://dp.la/search?utf8=%E2%9C%93&q=NASA; https://dp.la/search?q=NASA+African-American&utf8=%E2%9C%93.

9. "Primary Source Sets," https://dp.la/primary-source-sets.

10. For example, Amy Earhart, "Can Information Be Unfettered?: Race and the New Digital Humanities Canon," in *Debates in Digital Humanities*, ed. Matthew Gold (Minneapolis: University of Minnesota Press, 2012), 309–318; Sarah Potvin, "A Catalyst for Social Activism: The Digital Black Bibliographic Project," 2017 Texas Conference on Digital Humanities, https://conferences.tdl.org/tcdl/index.php/TCDL/TCDL2016/paper/viewFile/946/402.

11. "Digital Archives," http://www.blackpast.org/digital-archives. Its definition of an archive may be flexible in places.

12. https://webfiles.uci.edu/mcbrown/display/faces.html.

13. https://www.umbrasearch.org/pages/about.

14. "Information Center: About Black Studies Center," http://bsc.chadwyck.com/infoCenter/infoCenter.do?page=about.

15. David A. Graham, "Rumsfeld's Knowns and Unknowns: The Intellectual History of a Quip," *theAtlantic.com*, March 27, 2014, https://www.theatlantic.com/politics/archive/2014/03/rumsfelds-knowns-and-unknowns-the-intellectual-history-of-a-quip/359719/.

16. Matt Delmont, "Black Newspaper Resources Online," March 8, 2016, http://black-quotidian.com/anvc/black-quotidian/black-newspaper-resources-online; James A. Cannavino Library, Marist College, "Historical African American Newspapers Available Online: Home," http://libguides.marist.edu/c.php?g=87271&p=563206; Black Press Research Collective, http://blackpressresearchcollective.org/resources/scholarship-archives.

17. "Negroes Who Help Conquer Space: Over 1,000 Negroes Are in Satellite, Missile Field," *Ebony*, May 1958, 19–26.

18. Richard Paul interview, February 24, 2017.

19. Shanee' Yvette Murrain et al., "Radical Partnerships: Taking New Paths in Black Collections," *Archival Outlook* (January/February 2017), http://www.bluetoad.com/publication/?i=376049&article_id=2686571.

20. http://scalar.usc.edu/.

21. For example, for his *The Nicest Kids in Town: American Bandstand, Rock 'n' Roll, and the Struggle for Civil Rights in 1950s Philadelphia* (Berkeley: University of California Press, 2012), Matthew F. Delmont created http://nicestkids.com/nehvectors/nicest-kids/index.

22. Danielle Rios, Dianne Bohach, Jennifer Lam, and Bobbi deMontigny, "America and the Race to the Moon," Digital Public Library of America, October 2015, http://dp.la/exhibitions/exhibits/show/race-to-the-moon. See also, e.g., "Millican 'Riot,' 1868," http://millican.omeka.net/.

23. T. Harry Williams Center for Oral History, Louisiana State University, "The Costs of Doing Oral History," n.d., http://lib.lsu.edu/sites/all/files/oralhistory/resources/Oral_History_Budget.pdf.

24. Douglas A. Boyd, "Designing an Oral History Project: Initial Questions to Ask Yourself," in *Oral History in the Digital Age*, edited by Doug Boyd, Steve Cohen, Brad

Rakerd, and Dean Rehberger (Washington, DC: Institute of Museum and Library Services, 2012), http://ohda.matrix.msu.edu/2012/06/designing-an-oral-history-project/.

25. http://www.blackmuseums.org/.

26. Murrain et al., "Radical Partnerships."

27. Ibid.

28. Daren C. Brabham, "The Myth of Amateur Crowds," *Information, Communication & Society* 15, no. 3 (2012): 394–410, DOI: 10.1080/1369118X.2011.641991; Mike Bulajewski, "Crowdsourcing for Capitalism," June 3, 2012, http://www.metareader.org/post/crowdsourcing-for-capitalism.html.

29. Benjamin Vershbow, "Crowdsourcing Culinary History at the New York Public Library April 1, 2011–December 31, 2012, National Endowment for the Humanities Final White Paper Report Grant HD-51301–11," March 31, 2013, https://securegrants.neh.gov/PublicQuery/main.aspx?f=1&gn=HD-51301–11; New York Public Library, "What's on the Menu–About," http://menus.nypl.org/about.

30. National Archives, "Citizen Archivist Dashboard," November 17, 2016, https://www.archives.gov/citizen-archivist; Danielle Kim, "Crowdsourcing to Preserve Our Nation's History," Data-Smart City Solutions, October 28, 2014, http://datasmart.ash.harvard.edu/news/article/crowdsourcing-to-preserve-our-nations-history-553.

31. http://www.pvamu.edu/cresse/.

32. "NSBE's Aerospace Systems Conference Soars in Year Four," *Career Engineer* (Fall 2016): 50, 53.

33. http://www.ntaonline.org/history.html; Shetterly, *Hidden Figures*, 197.

34. T. Harry Williams Center for Oral History, Louisiana State University, "Costs of Doing Oral History."

35. Cindy George, "Houston Voices Join African-American Archive," *Houston Chronicle*, February 28, 2017, A1, 11.

36. David H. Guston, "Societal Dimensions Research in the National Nanotechnology Initiative," Consortium for Science, Policy & Outcomes, Arizona State University, May 2010, http://cspo.org/legacy/library/100701F7WL_lib_CSPOReportGuston.pdf.

37. Bonnie Malkin, "Hidden Figures No More: Female NASA Staff to Be Immortalized in Lego," *theguardian.com*, February 28, 2017, https://www.theguardian.com/lifeandstyle/2017/mar/01/hidden-figures-no-more-female-nasa-staff-to-be-immortalised-in-lego.

CONTRIBUTORS

P. J. Blount is a postdoctoral researcher in the Faculty of Law, Economics, and Finance at the University of Luxembourg and adjunct professor in the LL.M. in air and space law at the University of Mississippi School of Law.

Jonathan Coopersmith is professor of history at Texas A&M University.

Matthew L. Downs is chair of the Department of Social and Behavioral Sciences and associate professor of history at the University of Mobile.

Eric Fenrich is instructor of history at the University of California, Santa Barbara.

Cathleen Lewis is curator in the Department of Space History at the Smithsonian Institution's National Air and Space Museum.

Cyrus C. M. Mody is professor and chair in the History of Science, Technology, and Innovation at Maastricht University.

David Miguel Molina is lecturer in communication at the University of Pittsburgh and a doctoral candidate in rhetoric and public culture at Northwestern University.

Brian C. Odom is the center historian at NASA's Marshall Space Flight Center in Huntsville, Alabama.

Brenda Plummer is professor of history and Afro-American studies at the University of Wisconsin, Madison.

Christina K. Roberts is in her third year of the master's history program at the University of Nevada, Reno, focusing on the field of twentieth-century Soviet/Russian history as well as US and digital history.

Keith Snedegar is professor of history at Utah Valley University.

Stephen P. Waring is chair of the Department of History at the University of Alabama, Huntsville.

Margaret A. Weitekamp is curator in the Department of Space History at the Smithsonian Institution's National Air and Space Museum.

INDEX

Printed in the United States
By Bookmasters